M208 Pure

The Open University

GTA2

Groups and subgroups

The Open University, Walton Hall, Milton Keynes, MK7 6AA.

First published 2006. Reprinted with amendments 2007.

Edited, designed and typeset by The Open University, using the Open University TeX System.

Printed and bound in the United Kingdom by Hobbs the Printers Limited, Brunel Road, Totton, Hampshire SO40 3WX.

ISBN 0 7492 0216 5

1.2

Contents

Introduction

This unit builds on Unit GTA1 and introduces more of the basic ideas of group theory.

In Section 1 we define a *subgroup* of a group—a subset of elements that itself forms a group. In particular, we identify subgroups of symmetry groups by modifying figures.

In Section 2 we continue the search for subgroups by finding subgroups in which each element can be obtained, or *generated*, by repeatedly composing a single element with itself. This leads to a particular sort of group, a *cyclic group*, where the whole group is generated by a single element.

In Section 3, the video section, we formalise an observation made in Section 2: that two cyclic groups of the same order are 'essentially the same' group. This leads to the idea of *isomorphic groups*, which are groups with the same structure.

In Section 4 we investigate groups and subgroups that are based on modular arithmetics. We find that every additive modular arithmetic is a cyclic group and we identify the subgroups (all cyclic) of each additive modular arithmetic group. The properties of multiplicative modular arithmetics are not so straightforward to describe, but we find subsets of the set $\{0, 1, \ldots, n-1\}$ that do form groups under multiplication modulo n.

Study guide

We suggest that you study the sections in the natural order, but you can watch the video programme at any time before studying Section 4.

Sections 1 and 2 are fairly long sections that introduce important concepts. Section 1 is straightforward and does not contain many new ideas, although it contains a lot of exercises. However, Section 2, the audio section, introduces some abstract ideas and requires detailed study. You will probably need to spend more than one study session on each of Sections 1 and 2.

Section 3, the video section, introduces ideas which will be developed later in this block and in Group Theory Block B.

Section 4 is fairly short and may be read quickly.

1   Groups and subgroups

After working through this section, you should be able to:

(a) understand that a subset of a group may itself be a group;

(b) use the three subgroup properties to determine whether $(H, \circ)$ is a *subgroup* of a group $(G, \circ)$, where H is a subset of the set G;

(c) understand the differences between the group axioms G2 and G3 and the corresponding subgroup properties SG2 and SG3;

(d) find subgroups of a symmetry group $S(F)$ by adding some features to the figure F.

1.1   What is a subgroup?

In Unit GTA1 we explored the idea of a group, using examples from a variety of settings, including symmetries of a figure. We found the group table of the symmetry group $(S(\triangle), \circ)$ of an equilateral triangle:

See Unit GTA1, Subsection 5.2.

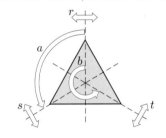

$\circ$	e	a	b	r	s	t
e	e	a	b	r	s	t
a	a	b	e	t	r	s
b	b	e	a	s	t	r
r	r	s	t	e	a	b
s	s	t	r	b	e	a
t	t	r	s	a	b	e

If we concentrate on the intersections of the rows and columns labelled by the symmetries e, a and b, then we obtain the subtable shown below.

The subtable is obtained by deleting the rows and columns labelled by the symmetries other than e, a and b.

$\circ$	e	a	b	r	s	t
e	e	a	b	r	s	t
a	a	b	e	t	r	s
b	b	e	a	s	t	r
r	r	s	t	e	a	b
s	s	t	r	b	e	a
t	t	r	s	a	b	e

$\circ$	e	a	b
e	e	a	b
a	a	b	e
b	b	e	a

This is the group table of $(S^+(\triangle), \circ)$, the group of direct symmetries of an equilateral triangle.

We saw in Unit GTA1 that the direct symmetries of any figure form a group under composition.

The two groups $(S(\triangle), \circ)$ and $(S^+(\triangle), \circ)$ are related: the set $S^+(\triangle)$ is a *subset* of $S(\triangle)$, and the set $S^+(\triangle)$ forms a group under the *same* binary operation $\circ$ as $(S(\triangle), \circ)$. We say that $(S^+(\triangle), \circ)$ is a *subgroup* of $(S(\triangle), \circ)$.

Definition   A **subgroup** of a group $(G, \circ)$ is a group $(H, \circ)$, where H is a subset of G.

Remark   It is part of the definition of a subgroup that the subgroup has the *same* binary operation as the parent group.

5

Some other groups that we have considered are given in the following list, arranged in pairs. In each pair, the underlying set of the second group is a subset of the underlying set of the first group, and the group operations are the same. Thus, in each pair the second group is a subgroup of the first.

See the Unit GTA1 subsections and exercises given below.

$(S(\square), \circ)$	the symmetries of a square,	Subsection 1.3
$(S^+(\square), \circ)$	the direct symmetries of a square;	Subsection 1.5
$(\mathbb{C}, +)$	the complex numbers under addition,	Exercise 3.8
$(\mathbb{R}, +)$	the real numbers under addition;	Exercise 3.1(a)
$(\mathbb{R}, +)$	the real numbers under addition,	Exercise 3.1(a)
$(\mathbb{Z}, +)$	the integers under addition;	See Subsection 3.2, Frame 3.
$(\mathbb{R}^*, \times)$	the non-zero real numbers under multiplication,	See Subsection 3.2, Frame 5.
$(\mathbb{Q}^*, \times)$	the non-zero rational numbers under multiplication.	Exercise 3.1(c)

Consider also the following pair of groups:

$(\mathbb{R}, +)$	the real numbers under addition,
$(\mathbb{R}^*, \times)$	the non-zero real numbers under multiplication.

In this case the second group is *not* a subgroup of the first because, although $\mathbb{R}^*$ is a subset of $\mathbb{R}$, the binary operations of the two groups are different.

Before we give more examples of subgroups of groups, we recall the formal definition of a group.

Definition   Let G be a set and $\circ$ be a binary operation defined on G. Then $(G, \circ)$ is a **group** if the following four axioms hold.

See Unit GTA1, Section 3, Frame 1.

G1 CLOSURE   For all $g_1, g_2 \in G$,

$$g_1 \circ g_2 \in G.$$

G2 IDENTITY   There exists an identity element $e \in G$ such that, for all $g \in G$,

$$g \circ e = g = e \circ g.$$

G3 INVERSES   For each $g \in G$, there exists an inverse element $g^{-1} \in G$ such that

$$g \circ g^{-1} = e = g^{-1} \circ g.$$

G4 ASSOCIATIVITY   For all $g_1, g_2, g_3 \in G$,

$$g_1 \circ (g_2 \circ g_3) = (g_1 \circ g_2) \circ g_3.$$

We also recall the strategy for determining whether a given set G and binary operation $\circ$ is a group.

See Unit GTA1, Section 3, Frame 2.

Strategy 1.1   To determine whether $(G, \circ)$ is a group,

GUESS behaviour, ... CHECK definition.

To show that $(G, \circ)$ is a group, show that EACH of the axioms G1, G2, G3 and G4 holds.

To show that $(G, \circ)$ is not a group, show that ANY ONE of the axioms G1, G2, G3 or G4 fails; that is:

	show that $\circ$ is not closed on G,	Axiom G1 fails.
OR	show that there is no identity element in G,	Axiom G2 fails.
OR	find one element in G with no inverse in G,	Axiom G3 fails.
OR	show that $\circ$ is not associative.	Axiom G4 fails.

Consider again the group table for $(S(\triangle), \circ)$. Let us determine whether the subset $\{e, s\}$ of $S(\triangle)$ forms a subgroup. We begin by finding a Cayley table for $\{e, s\}$ under $\circ$:

$\circ$	e	a	b	r	s	t
e	e	a	b	r	s	t
a	a	b	e	t	r	s
b	b	e	a	s	t	r
r	r	s	t	e	a	b
s	s	t	r	b	e	a
t	t	r	s	a	b	e

$\circ$	e	s
e	e	s
s	s	e

We apply Strategy 1.1 to determine whether $\{e, s\}$ is a group under $\circ$.

G1 CLOSURE   Only the two elements e and s occur in the body of the subtable, so $\{e, s\}$ is closed under the binary operation $\circ$.

G2 IDENTITY   We see that $\{e, s\}$ contains an identity element, e.

G3 INVERSES   We see that $e \circ e = s \circ s = e$.

Thus both elements have an inverse; in fact, they are both self-inverse.

G4 ASSOCIATIVITY   Composition of symmetries is associative.

Hence $(\{e, s\}, \circ)$ is a group. The set $\{e, s\}$ is a subset of the set $S(\triangle)$, so $(\{e, s\}, \circ)$ is a subgroup of $(S(\triangle), \circ)$.

Now consider the subset $\{e, b, r\}$ of $S(\triangle)$:

$\circ$	e	a	b	r	s	t
e	e	a	b	r	s	t
a	a	b	e	t	r	s
b	b	e	a	s	t	r
r	r	s	t	e	a	b
s	s	t	r	b	e	a
t	t	r	s	a	b	e

$\circ$	e	b	r
e	e	b	r
b	b	a	s
r	r	t	e

This subtable contains elements other than e, b and r, so $\{e, b, r\}$ is not closed under the operation $\circ$. Thus G1 CLOSURE fails and the set does not form a group. Hence the set $\{e, b, r\}$ under the operation $\circ$ is not a subgroup of $(S(\triangle), \circ)$.

Before proceeding with our discussion of subgroups, we deal with a rather subtle issue that arises in this context. We know that any group $(G, \circ)$ contains a unique identity element e. That is, there is a unique element $e \in G$ such that

$$g \circ e = e \circ g = g \quad \text{for all } g \in G.$$

Might there exist a subgroup H of G in which an element other than e is the identity element? In other words, might there be a subgroup H, and an element $e_H \in H$ not equal to e, with the property that

$$h \circ e_H = e_H \circ h = h \quad \text{for all } h \in H?$$

This is *not* possible. In any group $(G, \circ)$, the identity element e is the only element that satisfies the equation $x \circ x = x$. (To see this, compose both sides of this equation on the right by x^{-1}: we obtain $x = e$.) If the identity e_H of a subgroup H were different from the identity e of G, then e_H would be a second element of G satisfying $x \circ x = x$.

So the identity of a subgroup $(H, \circ)$ of a group $(G, \circ)$ must be the same as the identity of G. We can also prove that the inverse of each element in H must be the same as its inverse in G, but we omit the details here.

1.2 Checking for subgroups

In Subsection 1.1 we showed that $(\{e, s\}, \circ)$ is a subgroup of $(S(\triangle), \circ)$ by checking each group axiom for $(\{e, s\}, \circ)$. But it was not really necessary to check all the details that we considered since some properties hold in $(\{e, s\}, \circ)$ simply because they hold in $(S(\triangle), \circ)$. For example, we already know that $\circ$ is associative on $S(\triangle)$, so it must be associative on the subset $\{e, s\}$.

In general, given a group $(G, \circ)$ and a subset H of G, we can use our knowledge that $(G, \circ)$ is a group when we consider whether the four group axioms hold for H under $\circ$.

Let us look at each of the four axioms in turn to see what we need to check if we wish to prove that a subset H of a group $(G, \circ)$ is a subgroup.

G1 CLOSURE

> We want to show that, for all elements $h_1, h_2 \in H$, the composite $h_1 \circ h_2 \in H$.
>
> We need to check this in the usual way.

G2 IDENTITY

> We want to show that H contains an identity element.
>
> We know that G contains an identity element e. Thus $eh = he = h$ for all h in H.
>
> We need to check only that the identity element e of G belongs to H.

G3 INVERSES

> We want to show that, for each element $h \in H$, there is an inverse element $h^{-1} \in H$.
>
> Again, our knowledge of $(G, \circ)$ helps. We know that h^{-1} must exist in G, and that it is unique.
>
> We need to check only that, for each $h \in H$, the element h^{-1} belongs to H.

G4 ASSOCIATIVITY We want to show that, for all $h_1, h_2, h_3 \in H$,

$$h_1 \circ (h_2 \circ h_3) = (h_1 \circ h_2) \circ h_3.$$

This is the easiest of all! Associativity is a property of the operation defined on the parent set G. We know that, for all $g_1, g_2, g_3 \in G$,

$$g_1 \circ (g_2 \circ g_3) = (g_1 \circ g_2) \circ g_3$$

and H is a subset of G, so this result must be true in the particular case where the elements g_1, g_2 and g_3 belong to the subset H of G. Associativity in H is inherited from G.

We do not need to check associativity for a subgroup.

We formalise the discussion above as a theorem.

Theorem 1.1   Let $(G, \circ)$ be a group with identity e and let H be a subset of G. Then $(H, \circ)$ is a **subgroup** of $(G, \circ)$ if and only if the following three properties hold.

SG1 CLOSURE   For all $h_1, h_2 \in H$, the composite $h_1 \circ h_2 \in H$.

SG2 IDENTITY   The identity element $e \in H$.

SG3 INVERSES   For each $h \in H$, the inverse element $h^{-1} \in H$.

The 'if' part of the theorem follows from the discussion above. For the 'only if' part, property SG1 follows from the fact that a subgroup, being a group, must be closed, and properties SG2 and SG3 follow from the discussion at the end of Subsection 1.1.

Remark   We shall refer to SG1, SG2 and SG3 as *the three subgroup properties*. Although we have given them the same names as the group axioms, only SG1 involves the same ideas as the corresponding group axiom. The properties SG2 and SG3 require us to check only that certain elements (the identity of G and the inverses of elements of H) belong to H. There is no need to check that these elements have the defining properties of an identity or inverse since these properties follow from the axioms of G.

Every group with more than one element has at least two subgroups: the group $(G, \circ)$ itself, and the so-called **trivial subgroup** $(\{e\}, \circ)$ consisting of the identity element alone. A subgroup other than the whole group $(G, \circ)$ is called a **proper subgroup**.

Particular examples of the trivial subgroup are:
$(\{e\}, \circ)$, a subgroup of $(S(\triangle), \circ)$,
$(\{0\}, +)$, a subgroup of $(\mathbb{Z}, +)$,
$(\{1\}, \times)$, a subgroup of $(\mathbb{R}^*, \times)$.

The following strategy is similar to our strategy for groups.

Strategy 1.2   To determine whether $(H, \circ)$ is a subgroup of $(G, \circ)$, where $H \subseteq G$,

GUESS behaviour, ... CHECK definition.

To show that $(H, \circ)$ is a subgroup, show that EACH of the properties SG1, SG2 and SG3 holds.

To show that $(H, \circ)$ is not a subgroup, show that ANY ONE of the properties SG1, SG2 or SG3 fails; that is,

     show that $\circ$ is not closed on H,
OR show that $e \notin H$,
OR find one element $h \in H$ for which $h^{-1} \notin H$.

If $H \nsubseteq G$, then $(H, \circ)$ cannot be a subgroup of $(G, \circ)$.

Property SG1 fails.
Property SG2 fails.
Property SG3 fails.

The following examples illustrate the use of Strategy 1.2. First we look at subgroups of the symmetry group $(S(\square), \circ)$.

Example 1.1

(a) Show that $(\{e, b, r, t\}, \circ)$ is a subgroup of $(S(\square), \circ)$.

(b) Show that $(\{e, r, s, t\}, \circ)$ is not a subgroup of $(S(\square), \circ)$.

Solution

(a) We have $\{e, b, r, t\} \subseteq S(\square)$, and the binary operation $\circ$ is the same on each set.

We check the subgroup properties SG1, SG2 and SG3 in turn. We shall do this from the Cayley table for $(\{e, b, r, t\}, \circ)$, which can be extracted from the Cayley table for $S(\square)$ (in the margin) as follows.

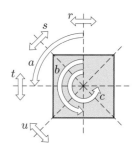

$\circ$	e	b	r	t
e	e	b	r	t
b	b	e	t	r
r	r	t	e	b
t	t	r	b	e

$\circ$	e	a	b	c	r	s	t	u
e	e	a	b	c	r	s	t	u
a	a	b	c	e	s	t	u	r
b	b	c	e	a	t	u	r	s
c	c	e	a	b	u	r	s	t
r	r	u	t	s	e	c	b	a
s	s	r	u	t	a	e	c	b
t	t	s	r	u	b	a	e	c
u	u	t	s	r	c	b	a	e

SG1 CLOSURE   No new elements are needed to complete the Cayley table, so $\{e, b, r, t\}$ is closed under composition.

SG2 IDENTITY   The identity in $S(\square)$ is e, and $e \in \{e, b, r, t\}$.

SG3 INVERSES   We see that the elements e, b, r and t are all self-inverse, so $\{e, b, r, t\}$ contains the inverse of each of its elements.

Hence $(\{e, b, r, t\}, \circ)$ satisfies the three subgroup properties, and so is a subgroup of $(S(\square), \circ)$.

(b) Here $\{e, r, s, t\} \subseteq S(\square)$, and the binary operation $\circ$ is the same on each set.

However, $r \circ t = b \notin \{e, r, s, t\}$, so property SG1 fails.

Hence $(\{e, r, s, t\}, \circ)$ is not a subgroup of $(S(\square), \circ)$. ∎

Exercise 1.1 Show that $(\{e, b, s, u\}, \circ)$ is a subgroup of $(S(\square), \circ)$.

Next we look at some infinite groups. The next example involves a finite subgroup of an infinite group. We proceed in a similar manner to Example 1.1.

Example 1.2   Show that $(\{1, -1, i, -i\}, \times)$ is a subgroup of $(\mathbb{C}^*, \times)$.

Solution   We have $\{1, -1, i, -i\} \subseteq \mathbb{C}^*$, and the binary operation $\times$ is the same on each set.

$\mathbb{C}^*$ is the set of non-zero complex numbers $\mathbb{C} - \{0\}$.

We check the subgroup properties SG1, SG2 and SG3 in turn. We do this from the Cayley table for $(\{1, -1, i, -i\}, \times)$, which is as follows.

$\circ$	1	-1	i	$-i$
1	1	-1	i	$-i$
-1	-1	1	$-i$	i
i	i	$-i$	-1	1
$-i$	$-i$	i	1	-1

SG1 CLOSURE   No new elements are needed to complete the Cayley table, so $\{1, -1, i, -i\}$ is closed under composition.

SG2 IDENTITY   The identity in $\mathbb{C}^*$ is 1, and $1 \in \{1, -1, i, -i\}$.

SG3 INVERSES   From the table, we see that the elements 1 and -1 are self-inverse, and i and $-i$ are inverses of each other, so $\{1, -1, i, -i\}$ contains the inverse of each of its elements.

Hence $(\{1, -1, i, -i\}, \times)$ satisfies the three subgroup properties, and so is a subgroup of $(\mathbb{C}^*, \times)$. ∎

The rest of the examples in this subsection involve infinite subgroups of infinite groups. In these cases we cannot use a Cayley table and have to argue using general elements of the groups.

Example 1.3   Show that $(\mathbb{R}^+, \times)$ is a subgroup of $(\mathbb{R}^*, \times)$.

Solution   Here $\mathbb{R}^+ \subseteq \mathbb{R}^*$, and the binary operation $\times$ is the same on each set.

$\mathbb{R}^+$ is the set of positive real numbers. $\mathbb{R}^*$ is the set of non-zero real numbers, $\mathbb{R} - \{0\}$.

We check the subgroup properties SG1, SG2 and SG3 in turn.

SG1 CLOSURE   Let $x, y \in \mathbb{R}^+$; then $x, y \in \mathbb{R}$, $x > 0$ and $y > 0$. It follows that $xy \in \mathbb{R}$ and $xy > 0$, so $xy \in \mathbb{R}^+$.

Thus $\mathbb{R}^+$ is closed under $\times$.

SG2 IDENTITY   The identity element in $(\mathbb{R}^*, \times)$ is 1 and $1 > 0$, so $1 \in \mathbb{R}^+$.

Thus $\mathbb{R}^+$ contains the identity element.

SG3 INVERSES   For $x \in \mathbb{R}^*$, the inverse is $1/x$. Now if $x \in \mathbb{R}^+$, then $x > 0$ and $1/x > 0$, so $1/x \in \mathbb{R}^+$.

Thus $\mathbb{R}^+$ contains the inverse of each of its elements.

Hence $(\mathbb{R}^+, \times)$ satisfies the subgroup properties SG1, SG2 and SG3, and so is a subgroup of $(\mathbb{R}^*, \times)$. ∎

Example 1.4   Show the following.

(a) $(\mathbb{Z}^+, +)$ is not a subgroup of $(\mathbb{Z}, +)$.

(b) $(4\mathbb{Z}, +)$ is not a subgroup of $(3\mathbb{Z}, +)$.

$\mathbb{Z}^+$ is the set of positive integers.

Here $(3\mathbb{Z}, +)$ and $(4\mathbb{Z}, +)$ are groups. In general, $(n\mathbb{Z}, +)$ is a group for any $n \in \mathbb{N}$: the proof is analogous to the solution to Exercise 3.9 in Unit GTA1.

Solution

(a) We have $\mathbb{Z}^+ \subseteq \mathbb{Z}$, and the binary operation $+$ is the same on each set. However, the identity in $(\mathbb{Z}, +)$ is 0, but $0 \notin \mathbb{Z}^+$, so property SG2 fails.

Hence $(\mathbb{Z}^+, +)$ is not a subgroup of $(\mathbb{Z}, +)$.

In fact, $(\mathbb{Z}^+, +)$ is not even a group as it has no identity element.

(b) The sets $4\mathbb{Z}$ and $3\mathbb{Z}$ are the sets of multiples of 4 and 3 respectively:

$$4\mathbb{Z} = \{\ldots, -4k, \ldots, -8, -4, 0, 4, 8, \ldots, 4k, \ldots\},$$
$$3\mathbb{Z} = \{\ldots, -3k, \ldots, -6, -3, 0, 3, 6, \ldots, 3k, \ldots\}.$$

Now $4 \in 4\mathbb{Z}$, but $4 \notin 3\mathbb{Z}$, so $4\mathbb{Z}$ is not a subset of $3\mathbb{Z}$.

Hence $(4\mathbb{Z}, +)$ is not a subgroup of $(3\mathbb{Z}, +)$. ∎

Exercise 1.2

(a) Show that $(3\mathbb{Z}, +)$ is a subgroup of $(\mathbb{Z}, +)$.

(b) Show that $(6\mathbb{Z}, +)$ is a subgroup of $(2\mathbb{Z}, +)$.

Exercise 1.3   Show that $(\mathbb{Q}^*, \times)$ is not a subgroup of $(\mathbb{R}^+, \times)$.

$\mathbb{Q}^*$ is the set of non-zero rational numbers, $\mathbb{Q} - \{0\}$.

The following two exercises should familiarise you further with using Strategy 1.2.

Exercise 1.4 Show that the following subsets of $\mathbb{Z}_{12}$, together with the operation $+_{12}$, are subgroups of $(\mathbb{Z}_{12}, +_{12})$.

(a) $H_1 = \{0, 3, 6, 9\}$ (b) $H_2 = \{0, 4, 8\}$

Exercise 1.5 In each of the following cases, H is a subset of G, but $(H, \circ)$ is not a subgroup of $(G, \circ)$. Explain why not.

(a) $(G, \circ) = (S(\square), \circ)$ and $(H, \circ) = (\{e, a, c\}, \circ)$.

(b) $(G, \circ) = (\mathbb{Z}_5^*, \times_5)$ and $(H, \circ) = (\{2, 3, 4\}, \times_5)$.

(c) $(G, \circ) = (\mathbb{R}^*, \times)$ and $(H, \circ) = (\mathbb{Z}^*, \times)$.

An unfamiliar operation

The following example is designed to illustrate the differences between using the group axioms (in part (a)) and the three subgroup properties (in parts (b) and (c)).

In this example, the operation is probably unfamiliar to you and the identity and the inverses are not obvious. Also, the group is non-Abelian, so we have to check the 'two-sided' properties of the identity and inverses. That is, for the identity we must check that both $g \circ e$ and $e \circ g$ are equal to g, and for inverses we must check that both $g \circ g^{-1}$ and $g^{-1} \circ g$ are equal to e.

Example 1.5   Let X be the subset of $\mathbb{R}^2$ consisting of all the points not on the y-axis; that is,

$$X = \{(a, b) \in \mathbb{R}^2 : a \neq 0\}.$$

Let $*$ be the binary operation on X defined by

$$(a, b) * (c, d) = (ac, ad + b).$$

(a) Show that $(X, *)$ is a group.

For each of the following subsets of X, determine whether the subset, together with the binary operation $*$, is a subgroup of $(X, *)$.

(b) $A = \{(a, b) \in X : a = 1\}$

(c) $B = \{(a, b) \in X : b = 1\}$

Solution

(a) **Comments**

We follow Strategy 1.1 for groups.

We consider two typical elements of X.

We write down the condition for membership of X.

We write down the product, and use closure of $+$ and $\times$ on $\mathbb{R}$.

We check the condition for membership of X.

This proves that axiom G1 holds.

For example,
$$(2, 5) * (4, 3)$$
$$= (2 \times 4, 2 \times 3 + 5) = (8, 11),$$
and
$$(-2, \pi) * (\tfrac{1}{4}, 0)$$
$$= ((-2) \times \tfrac{1}{4}, (-2) \times 0 + \pi)$$
$$= (-\tfrac{1}{2}, \pi).$$

Proof

We show that $(X, *)$ satisfies the four group axioms.

G1 CLOSURE

Let $(a, b), (c, d) \in X$;

then $a, b, c, d \in \mathbb{R}$, and $a \neq 0$ and $c \neq 0$.

By definition,
$$(a, b) * (c, d) = (ac, ad + b) \in \mathbb{R}^2;$$
also, $ac \neq 0$ because $a \neq 0$ and $c \neq 0$, so
$$(a, b) * (c, d) \in X.$$

So X is closed under $*$.

An identity element is not obvious, so we assume that an identity (x, y) exists, and we try to find it.

G2 IDENTITY

Suppose that $(x, y) \in X$ is an identity in X.

We write down the condition that (x, y) acts as an identity on the right.

Then we must have, for each $(a, b) \in X$,
$$(a, b) * (x, y) = (a, b);$$
that is,
$$(ax, ay + b) = (a, b).$$

We equate the coordinates.

Comparing coordinates, we obtain
$$ax = a \quad \text{and} \quad ay + b = b.$$

We can solve these equations because $a \neq 0$.

Now $a \neq 0$, so we must have
$$x = 1 \quad \text{and} \quad y = 0;$$

If there is an identity, it must be $(1, 0)$.

so the element $(1, 0)$ acts as an identity on the right.

Now we check that $(1, 0)$ acts as an identity on the left.

Also,
$$(1, 0) * (a, b) = (1 \times a, 1 \times b + 0) = (a, b),$$
so $(1, 0)$ acts as an identity on the left.

We check that $(1, 0)$ belongs to X by noting that it is an element of $\mathbb{R}^2$ with a non-zero first coordinate (see the definition of X).

The element $(1, 0) \in X$ because $(1, 0)$ belongs to $\mathbb{R}^2$ and $1 \neq 0$.

We have shown that, for each $(a, b) \in X$,
$$(a, b) * (1, 0) = (a, b) = (1, 0) * (a, b).$$

This proves that axiom G2 holds.

so $(1, 0)$ is an identity in X.

G3 INVERSES

An inverse of a typical element (a, b) is not obvious, so we assume that an inverse (x, y) exists, and we try to find it.

Let $(a, b) \in X$; then $a, b \in \mathbb{R}$ and $a \neq 0$. Suppose that $(a, b)^{-1} = (x, y)$, where $(x, y) \in X$.

We write down the condition that (x, y) acts as an inverse on the right.

Then we must have
$$(a, b) * (x, y) = (1, 0);$$
that is,
$$(ax, ay + b) = (1, 0).$$

We equate the coordinates.

Comparing coordinates, we obtain
$$ax = 1 \quad \text{and} \quad ay + b = 0.$$

We can solve these equations because $a \neq 0$.

Now $a \neq 0$, so we must have
$$x = 1/a \quad \text{and} \quad y = -b/a,$$

If there is an inverse of (a, b), it must be $(1/a, -b/a)$.

so the element $(1/a, -b/a)$ acts as an inverse on the right.

Now we check that $(1/a, -b/a)$ acts as an inverse on the left.

Also,
$$(1/a, -b/a) * (a, b) = ((1/a) \times a, (1/a) \times b - b/a)$$
$$= (1, 0),$$
so $(1/a, -b/a)$ acts as an inverse on the left.

so the element $(1/a, -b/a)$ acts as an inverse on the right.

We check that $(1/a, -b/a)$ belongs to X by noting that it is an element of $\mathbb{R}^2$ with a non-zero first coordinate.

The element $(1/a, -b/a) \in X$ because $(1/a, -b/a) \in \mathbb{R}^2$ and $1/a \neq 0$.

We have shown that
$$(a, b) * (1/a, -b/a) = (1, 0)$$
and
$$(1/a, -b/a) * (a, b) = (1, 0),$$
so $(1/a, -b/a)$ is an inverse of (a, b).

This proves that axiom G3 holds.   Hence X contains an inverse of each of its elements.

G4 ASSOCIATIVITY

We consider three typical elements   Let $(a, b), (c, d), (e, f) \in X$.
of X.

We must show that
$$(a, b) * ((c, d) * (e, f)) = ((a, b) * (c, d)) * (e, f).$$

We expand each side separately.   First we consider the left-hand side:

Now we use associativity of addition
$$(a, b) * ((c, d) * (e, f)) = (a, b) * (ce, cf + d)$$
and multiplication on $\mathbb{R}$.
$$= (ace, acf + ad + b). \quad (1.1)$$

Now we consider the right-hand side:

Again we use associativity of addition
$$((a, b) * (c, d)) * (e, f) = (ac, ad + b) * (e, f)$$
and multiplication on $\mathbb{R}$.
$$= (ace, acf + ad + b). \quad (1.2)$$

This proves that axiom G4 holds.   Expressions (1.1) and (1.2) are the same, so $*$ is an associative operation on X.

This completes the proof.   Hence $(X, *)$ satisfies the four group axioms, so $(X, *)$ is a group.

(b) Comments

Proof

We follow Strategy 1.2 for subgroups.   Next, we consider $(A, *)$, where
$$A = \{(a, b) \in X : a = 1\} = \{(1, b) : b \in \mathbb{R}\}.$$

We notice that $(1, 0)$, the identity   We guess that $(A, *)$ is a subgroup of $(X, *)$.
in X, belongs to A, so $(A, *)$ *may* be a
subgroup.

To prove that our guess is correct, we   We check whether $(A, *)$ satisfies the three subgroup
must show that each of the properties   properties.
SG1, SG2 and SG3 holds.

SG1 CLOSURE

We consider two typical elements of A.   Let $(1, b)$ and $(1, d)$ be elements of A.

We write down the product, and check   By definition,
that it belongs to A.
$$(1, b) * (1, d) = (1 \times 1, 1 \times d + b) = (1, d + b);$$
this belongs to A because the first coordinate is 1.

This proves that property SG1 holds.   So A is closed under $*$.

SG2 IDENTITY

We check that the known   The identity in X is $(1, 0)$, and $(1, 0)$ belongs to A
identity $(1, 0)$ in X belongs to A.   because the first coordinate is 1.

This proves that property SG2 holds.   So A contains the identity.

SG3 INVERSES

We check that the known inverse   Let $(1, b) \in A$. Then
of $(1, b)$ in X belongs to A. We use
the result $(a, b)^{-1} = (1/a, -b/a)$
$$(1, b)^{-1} = (1/1, -b/1) = (1, -b);$$
with $a = 1$.   this belongs to A because the first coordinate is 1.

This proves that property SG3 holds.   So A contains the inverse of each of its elements.

This completes the proof.   Hence $(A, *)$ satisfies the three subgroup properties, so $(A, *)$ is a subgroup of $(X, *)$.

(c) **Comments**

To show that $(B, *)$ is not a subgroup, it is sufficient to show that ONE of the properties SG1, SG2 or SG3 fails.

In fact, SG1 and SG3 fail too.

Proof

Next, we consider $(B, *)$, where
$$B = \{(a, b) \in X : b = 1\} = \{(a, 1) : a \in \mathbb{R}^*\}.$$

We notice that $(1, 0)$, the identity in X, does not belong to B, so property SG2 fails.

Hence $(B, *)$ is not a subgroup of $(X, *)$. ∎

Exercise 1.6 Let X be the set
$$X = \{(a, b) \in \mathbb{R}^2 : ab \neq 0\}$$
and let $*$ be the binary operation on X defined by
$$(a, b) * (c, d) = (ac, bd).$$

(a) Show that $(X, *)$ is a group.

Determine which of the following subsets, together with the operation $*$, are subgroups of $(X, *)$.

(b) $A = \{(a, b) \in X : a = 1\} = \{(1, b) : b \in \mathbb{R}^*\}$

(c) $B = \{(a, b) \in X : a + b = 2\} = \{(a, 2 - a) : a \in \mathbb{R}^*,\ a \neq 2\}$

For example,
$$(-1, 4) * (3, 2)$$
$$= ((-1) \times 3, 4 \times 2)$$
$$= (-3, 8)$$
and
$$(\tfrac{1}{4}, -3) * (8, \sqrt{2})$$
$$= (\tfrac{1}{4} \times 8, -3 \times \sqrt{2})$$
$$= (2, -3\sqrt{2}).$$

1.3 Subgroups of symmetry groups

In Unit GTA1 we saw that the symmetries of any figure F form a group (under composition of functions) called the symmetry group of F, denoted by $S(F)$. We have also seen one particular subgroup of $S(F)$: the subgroup $S^+(F)$ of direct symmetries of F, which may or may not be equal to the whole symmetry group $S(F)$.

We omit the symbol ∘ from the expression for a symmetry group $(S(F), \circ)$ and write simply $S(F)$, whenever this will not cause confusion.

Another way of finding subgroups of a symmetry group $S(F)$ is to modify the figure F to restrict its symmetry. We could, for example, introduce a pattern of shapes into a square F to produce the modified square F' shown in Example 1.6. This new figure has the symmetry group $S(F')$; it consists of those symmetries of the square that leave the pattern of shapes unchanged and it is a proper subgroup $S(\square)$.

Like all figures, a figure with a pattern is just a subset of $\mathbb{R}^2$. The subset consists of all those points which lie within the blue shaded areas.

Example 1.6   Let F' be the the modified square shown below. Find a subgroup of $S(\square)$ by listing the symmetries of the figure F' .

F'

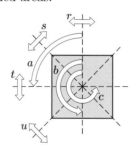

Solution   The effect of the modification is that the rotations through $\pi/2$ and $3\pi/2$ and the reflections s and u are no longer symmetries of the figure. So
$$S(F') = \{e, b, r, t\}$$
is a subgroup of $S(\square)$. ∎

Exercise 1.7 Find three subgroups of $S(\square)$ by listing the elements of the symmetry groups of each of the following modified squares.

(a)   (b)   (c)

We often use a simple pattern of line segments to restrict the symmetries of a figure. The following example uses a single diagonal line to restrict the symmetries of a square.

Example 1.7 Find a subgroup of $S(\square)$ by listing the symmetries of the following modified square.

Solution The effect of adding the diagonal line is that the rotations through $\pi/2$ and $3\pi/2$ and the reflections r and t are no longer symmetries of the figure. So

$$S(F') = \{e, b, s, u\}$$

is a subgroup of $S(\square)$. ∎

The above approach can be used to find subgroups of the symmetry groups of other figures.

Exercise 1.8 Let F be a regular hexagon. Describe geometrically the elements of the subgroup $S(F')$ of $S(F)$, where F' is the figure obtained by inscribing an equilateral triangle inside F as shown.

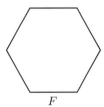 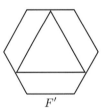

F $\qquad\qquad$ F'

Another way of finding a subgroup of a symmetry group $S(F)$ is to fix a particular vertex (or vertices). For example, the symmetries of a square that do not move the vertex at location 1 are e and s, so by fixing this vertex we obtain the subset $\{e, s\}$ of $S(\square)$. We saw that this subset is a subgroup of $S(\square)$ in Subsection 1.1.

Fixing a vertex of a figure F always yields a subgroup of $S(F)$ because fixing a vertex has the same effect as modifying the figure by adding a small dot at the vertex (differently sized dots if more than one vertex is to be fixed).

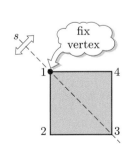

$S(\text{TET})$ revisited

We now use the method of fixing vertices to find some subgroups of $S(\text{TET})$, the group of symmetries of a regular tetrahedron.

We studied $S(\text{TET})$ in Unit GTA1, Section 5.

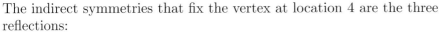

The direct symmetries of the tetrahedron that fix the vertex at location 4 are the three rotations through 0, $2\pi/3$ and $4\pi/3$ radians about the axis through this vertex and the centre of the opposite face:

$$\begin{pmatrix} 1 & 2 & 3 & 4 \\ 1 & 2 & 3 & 4 \end{pmatrix}, \quad \begin{pmatrix} 1 & 2 & 3 & 4 \\ 2 & 3 & 1 & 4 \end{pmatrix}, \quad \begin{pmatrix} 1 & 2 & 3 & 4 \\ 3 & 1 & 2 & 4 \end{pmatrix}.$$

The indirect symmetries that fix the vertex at location 4 are the three reflections:

$$\begin{pmatrix} 1 & 2 & 3 & 4 \\ 2 & 1 & 3 & 4 \end{pmatrix}, \quad \begin{pmatrix} 1 & 2 & 3 & 4 \\ 3 & 2 & 1 & 4 \end{pmatrix}, \quad \begin{pmatrix} 1 & 2 & 3 & 4 \\ 1 & 3 & 2 & 4 \end{pmatrix}.$$

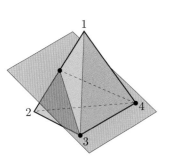 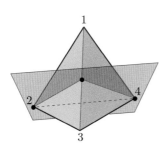 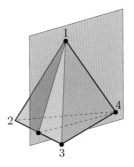

So the set of all symmetries that fix the vertex at location 4 is

$$V_4 = \left\{ \begin{pmatrix} 1 & 2 & 3 & 4 \\ 1 & 2 & 3 & 4 \end{pmatrix}, \begin{pmatrix} 1 & 2 & 3 & 4 \\ 2 & 3 & 1 & 4 \end{pmatrix}, \begin{pmatrix} 1 & 2 & 3 & 4 \\ 3 & 1 & 2 & 4 \end{pmatrix}, \right.$$
$$\left. \begin{pmatrix} 1 & 2 & 3 & 4 \\ 2 & 1 & 3 & 4 \end{pmatrix}, \begin{pmatrix} 1 & 2 & 3 & 4 \\ 3 & 2 & 1 & 4 \end{pmatrix}, \begin{pmatrix} 1 & 2 & 3 & 4 \\ 1 & 3 & 2 & 4 \end{pmatrix} \right\}.$$

Remark   The elements of V_4 look exactly like the elements of $S(\triangle)$, the symmetry group of an equilateral triangle, written as two-line symbols, but with an extra column (mapping 4 to 4) at the end. We defined V_4 to be the set of elements of $S(\text{TET})$ that fix the vertex at location 4, but we can also consider V_4 to be the set of symmetries of the triangular face with vertices at locations 1, 2 and 3. So V_4 has many features in common with $S(\triangle)$. Similarly, we can define subgroups V_1, V_2 and V_3 of $S(\text{TET})$, fixing the vertices at locations 1, 2 and 3, respectively, all having features similar to those of $S(\triangle)$.

We shall return to this idea of 'similar' groups in Section 3.

Exercise 1.9   Write down the two-line symbols for the symmetries of $S(\text{TET})$ that fix the vertex at location 3.

Summary

In this subsection we have seen various ways of finding subgroups of the symmetry group of a given figure.

Strategy 1.3   To find a subgroup of a given symmetry group of a figure, carry out *one* of the following.

* Modify the figure to restrict its symmetry, for example, by introducing a pattern of lines or shapes, and then determine which of the symmetries of the original figure are symmetries of the new figure.

* Find the symmetries of the figure that fix a particular vertex (or particular vertices).

Exercise 1.10 Let F be a triangular prism with equilateral triangles at its ends.

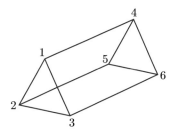

Find three subgroups of $S(F)$ by considering the following ways of restricting the symmetry.

(a) Add a vertical line to each of the triangular ends, as shown.

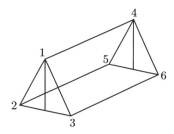

(b) Fill in one of the triangular ends, as shown.

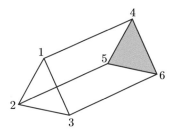

(c) Fix the vertices at locations 1 and 4.

Further exercises

Exercise 1.11 Prove the following statements using the three subgroup properties.

(a) $(11\mathbb{Z}, +)$ is a subgroup of $(\mathbb{Z}, +)$.

(b) $(\{10^k : k \in \mathbb{Z}\}, \times)$ is a subgroup of $(\mathbb{R}^*, \times)$.

(c) $(\{z \in \mathbb{C} : z = x + ix, \ x \in \mathbb{R}, \}, +)$ is a subgroup of $(\mathbb{C}, +)$.

(d) $(\{0, 3, 6, 9, 12\}, +_{15})$ is a subgroup of $(\mathbb{Z}_{15}, +_{15})$.

Exercise 1.12 Let $(X, *)$ be the group introduced in Example 1.5 and let

$$C = \{(a, b) \in X : b = 0\}.$$

Show that $(C, *)$ is a subgroup of $(X, *)$.

Exercise 1.13 Let X be the subset of $\mathbb{R}^2$ consisting of all points not on the x-axis, that is,

$$X = \{(x, y) \in \mathbb{R}^2 : y \neq 0\},$$

and let $*$ be the binary operation on X defined by

$$(a, b) * (c, d) = (ad + bc, bd).$$

You may assume that $(X, *)$ is a group, that the identity element of $(X, *)$ is $(0, 1)$ and that the inverse of (a, b) is $(-a/b^2, 1/b)$.

(a) Find $(-1, 3) * \left(\frac{1}{2}, 2\right)$ and $(0, 3) * (-1, 4)$.

(b) Let $A = \{(x, y) \in X : x = 0\} = \{(0, y) : y \in \mathbb{R}^*\}$.
    Show that $(A, *)$ is a subgroup of $(X, *)$.

(c) Let $B = \{(x, y) \in X : x = y\} = \{(x, x) : x \in \mathbb{R}^*\}$.
    Show that $(B, *)$ is not a subgroup of $(X, *)$.

Exercise 1.14 List the symmetries of each of the following modified regular polygons.

(a)   (b)   (c)

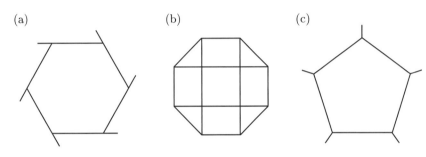

Exercise 1.15 List the symmetries of each of the following modified squares.

(a)   (b)   (c)

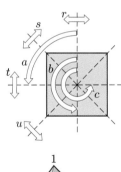

Exercise 1.16 Write down the elements of the subgroup V_{12} of $S(\text{TET})$ consisting of the symmetries of a tetrahedron that fix the *edge* joining the vertices originally at locations 1 and 2. (Fixing an edge of a figure F always yields a subgroup of $S(F)$ because fixing an edge has the same effect as adding a small distinguishing dot midway along the edge.)

An edge can be fixed without each point on the edge being fixed.

2 Cyclic groups

After working through this section, you should be able to:

(a) determine the subgroup generated by an element in a group;

(b) find the *order* of an element in a group;

(c) explain the meaning of the terms *cyclic group* and *generator*;

(d) find all the cyclic subgroups of a given finite group;

(e) understand that an element and its inverse generate the same cyclic subgroup.

2.1 What is a cyclic group?

In this section we introduce the idea of a *cyclic group*. This is a group in which all the elements can be obtained by repeatedly composing one element with itself; we say that this element *generates* the group. We shall see that forming cyclic groups gives us a source of subgroups of a group.

Certain important properties of a group are inherited by all its subgroups. For example, if a group is finite, then all its subgroups must also be finite; and if a group is Abelian, then all its subgroups must also be Abelian. We shall see that being cyclic is also an 'inherited property': if a group is cyclic, then all its subgroups must be cyclic.

We have already met several examples of cyclic groups. For example, the direct symmetries of a regular n-sided polygon (an n-gon) are all composites of a 'basic' rotation—a rotation through $2\pi/n$ about the centre.

Before you work through the audio, we should like you to try repeatedly composing an element with itself in two familiar groups. The following two exercises are discussed on the audio.

Exercise 2.1 In the group $S(\square)$, find the following composites.

(a) $a \circ a$
$a \circ a \circ a$
$a \circ a \circ a \circ a$
$a \circ a \circ a \circ a \circ a$

(b) $a^{-1} \circ a^{-1}$
$a^{-1} \circ a^{-1} \circ a^{-1}$
$a^{-1} \circ a^{-1} \circ a^{-1} \circ a^{-1}$
$a^{-1} \circ a^{-1} \circ a^{-1} \circ a^{-1} \circ a^{-1}$

(c) $b \circ b$
$b \circ b \circ b$

(d) $b^{-1} \circ b^{-1}$
$b^{-1} \circ b^{-1} \circ b^{-1}$

(e) What do you obtain if you repeatedly compose each of the elements r, s, t and u with itself?

Exercise 2.2 In the group $(\mathbb{Z}_6, +_6)$, find the following composites.

(a) $2 +_6 2$
$2 +_6 2 +_6 2$
$2 +_6 2 +_6 2 +_6 2$

(b) $4 +_6 4$
$4 +_6 4 +_6 4$
$4 +_6 4 +_6 4 +_6 4$

(c) $3 +_6 3$
$3 +_6 3 +_6 3$

(d) $1 +_6 1$
$1 +_6 1 +_6 1$
$1 +_6 1 +_6 1 +_6 1$
$1 +_6 1 +_6 1 +_6 1 +_6 1$

(e) $5 +_6 5$
$5 +_6 5 +_6 5$
$5 +_6 5 +_6 5 +_6 5$
$5 +_6 5 +_6 5 +_6 5 +_6 5$

Listen to the audio as you work through the frames.

Audio

3. Notation

multiplicative notation	additive notation
identity 1 or e	identity 0
$x \times x = x^2$	$x + x = 2x$
inverse x^{-1}	inverse $-x$
x^n	nx
x^{-n}	$-nx$
$x^s \times x^t = x^{s+t}$	$sx + tx = (s+t)x$

 *(cloud: powers)* *(cloud: multiples)*

$$\langle x \rangle = \{ x^k : k \in \mathbb{Z} \} \qquad \langle x \rangle = \{ kx : k \in \mathbb{Z} \}$$

4. Multiples of 2 in $(\mathbb{Z}, +)$

Let $(G, o) = (\mathbb{Z}, +)$; $x = 2$.

$$\langle 2 \rangle = \{ ..., -k2, -(k-1).2, ..., -2.2, -1.2, 0.2, 1.2, 2.2, ..., (k-1).2, k.2, ... \}$$

$$= \{ ..., -2k, -2(k-1), ..., -4, -2, 0, 2, 4, ..., 2(k-1), 2k, ... \}$$

(cloud: inverse of $2k$ is $-2k$)

(cloud: $x^{-1} = -2$)

1. Powers of an element x in a group (G, o)

$x^1 = x$ x^{-1} is inverse of x

$x^2 = x \circ x$ $x^{-2} = x^{-1} \circ x^{-1}$

$x^3 = x \circ x \circ x$ $x^{-3} = x^{-1} \circ x^{-1} \circ x^{-1}$

are all members of G, by group axioms.

Definition

$x^0 = e$

$x^n = x^{n-1} \circ x$

$x^{-n} = (x^{-1})^n$

 *(cloud: $n \in \mathbb{Z}^+$)*

*(cloud: set **generated** by x)*

$$\langle x \rangle = \{ x^k : k \in \mathbb{Z} \}$$

2. Powers of 2 in $(\mathbb{R}^*, \times)$

Let $(G, o) = (\mathbb{R}^*, \times)$; $x = 2$.

$$\langle 2 \rangle = \{ ..., 2^{-k}, 2^{-(k-1)}, ..., 2^{-2}, 2^{-1}, 2^0, 2^1, 2^2, ..., 2^{k-1}, 2^k, ... \}$$

$$= \{ ..., \frac{1}{2^k}, \frac{1}{2^{k-1}}, ..., \frac{1}{4}, \frac{1}{2}, 1, 2, 4, ..., 2^{k-1}, 2^k, ... \}$$

 *(cloud: inverse of 2^k is $\frac{1}{2^k} = \left(\frac{1}{2}\right)^k$)*

(cloud: $x^{-1} = \frac{1}{2}$)

21

5. Sets generated in S(□)

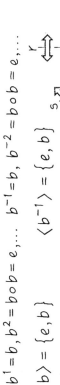

$a^1 = a$ $a^{-1} = c$

$a^2 = a \circ a = b$ $a^{-2} = c \circ c = b$

$a^3 = b \circ a = c$ $a^{-3} = b \circ c = a$

$a^4 = c \circ a = e$ $a^{-4} = a \circ c = e$

$a^5 = e \circ a = a$ $a^{-5} = e \circ c = c$

$a^4 = e$

$\langle a \rangle = \{e, a, b, c\}$ $\langle c \rangle = \{e, a, b, c\}$

$\qquad = \{e, a, a^2, a^3\}$ $\qquad = \{e, a^{-1}, a^{-2}, a^{-3}\}$

$b^1 = b, \; b^2 = b \circ b = e, \ldots$ $b^{-1} = b, \; b^{-2} = b \circ b = e, \ldots$

so $\langle b \rangle = \{e, b\}$ $\langle b^{-1} \rangle = \{e, b\}$

$\langle e \rangle = \{e\}$

$\langle r \rangle = \{e, r\}$ $\langle t \rangle = \{e, t\}$

$\langle s \rangle = \{e, s\}$ $\langle u \rangle = \{e, u\}$

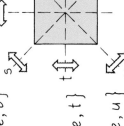

6. Sets generated in $\mathbb{Z}_6$

$\langle 0 \rangle = \{0\}$ $\langle 3 \rangle = \{0, 3\}$

$\langle 2 \rangle = \{0, 2, 4\}$ $\langle 1 \rangle = \{0, 1, 2, 3, 4, 5\}$

$\langle 4 \rangle = \{0, 4, 2\}$ $\langle 5 \rangle = \{0, 5, 4, 3, 2, 1\}$

7. Cycles

$S^+(\square)$

$e = a^4$, a, $a^{-1} = c$, $c = a^3$

$b = a^2 = c^2$

$\langle a \rangle = \langle c \rangle = \{e, a, b, c\}$

$\langle b \rangle = \{e, b\}$

$\langle e \rangle = \{e\}$

$\mathbb{Z}_6$

$-_6 1 = +_6 5 \quad (\mathrm{mod}\,6)$

$\langle 1 \rangle = \langle 5 \rangle = \mathbb{Z}_6$

$\langle 2 \rangle = \langle 4 \rangle = \{0, 2, 4\}$

$\langle 3 \rangle = \{0, 3\}$

$\langle 0 \rangle = \{0\}$

8. Properties of generated sets

- $\langle e \rangle = \{e\}$
- $\langle x^{-1} \rangle = \langle x \rangle$
- If x is self-inverse, i.e. $x^2 = e$, then $\langle x \rangle = \{e, x\}$

9. Exercise 2.3

Find the set generated by each element in

(a) $S(\square)$;   (b) $(\mathbb{Z}_5^*, \times_5)$;   (c) $S(\triangle)$.

13. Set generated by element x of order n

order of an element x

= number of elements in $\langle x \rangle$.

Theorem 2.1 If x has order n, then

$\langle x \rangle$ has exactly n distinct elements:

(multiplicative) $\langle x \rangle = \{e, x, x^2, \ldots, x^{n-1}\}$

or

(additive) $\langle x \rangle = \{0, x, 2x, \ldots, (n-1)x\}$.

> $x^n = e$

> $x^{n-1} = x^{-1}$

> $(n-1)x = -x$

> $\langle x \rangle$ is the set of *all powers of x*.

14. The subgroup $\langle x \rangle$

Theorem 2.2 Let $x \in G$;

then $\langle x \rangle$ is a subgroup of G.

Proof

SG1 CLOSURE $x^s \, x^t = x^{s+t} \in \langle x \rangle$;

SG2 IDENTITY $e = x^0$, so $e \in \langle x \rangle$;

SG3 INVERSES $x^s \, x^{-s} = e = x^{-s} \, x^s$,

so inverse of x^s is $x^{-s} \in \langle x \rangle$.

Hence $\langle x \rangle$ is a subgroup of G.

10. Elements of finite order

Definition Let $x \in (G, \circ)$.

If n is the *least* positive integer such that $x^n = e$,

then x has **order** n.

11. Orders of elements

$S(\square)$

element	order
e	1
a	4
c	4
b	2
r	2
s	2
t	2
u	2

$\mathbb{Z}_6$

element	order
0	1
2	3
4	3
3	2
1	6
5	6

> In $S(\square)$,
> $a^1 = a$, $a^2 = b$,
> $a^3 = c$, $a^4 = e$;
> so a has
> order 4.

> In $\mathbb{Z}_6$,
> $2 +_6 2 = 4$,
> $2 +_6 2 +_6 2 = 0$,
> so 2 has
> order 3.

12. Exercise 2.4

Find the order of each element in

(a) $S(\square)$; (b) $(\mathbb{Z}_5^*, \times_5)$; (c) $S(\triangle)$.

15. Cyclic subgroup

Definitions If $x \in G$, then

$\langle x \rangle$ is a **cyclic subgroup** of G;

$\langle x \rangle$ is **generated** by x;

x is a **generator** of $\langle x \rangle$.

- If x has order n, then $\langle x \rangle$ is a cyclic subgroup of order n.

16. Cyclic subgroups of $S(\square)$ and $\mathbb{Z}_6$.

$S(\square)$

subgroup	order
$\langle e \rangle = \{e\}$	1
$\langle a \rangle = S^+(\square)$	4
$\langle c \rangle = S^+(\square)$	4
$\langle b \rangle = \{e, b\}$	2
$\langle r \rangle = \{e, r\}$	2
$\langle s \rangle = \{e, s\}$	2
$\langle t \rangle = \{e, t\}$	2
$\langle u \rangle = \{e, u\}$	2

$\mathbb{Z}_6$

subgroup	order
$\langle 0 \rangle = \{0\}$	1
$\langle 2 \rangle = \{0, 2, 4\}$	3
$\langle 4 \rangle = \{0, 2, 4\}$	3
$\langle 3 \rangle = \{0, 3\}$	2
$\langle 1 \rangle = \mathbb{Z}_6$	6
$\langle 5 \rangle = \mathbb{Z}_6$	6

- no element generates $S(\square)$
- 1 or 5 generates $\mathbb{Z}_6$

17. Cyclic group

Definition If there exists $x \in G$ such that

$\langle x \rangle = G$, then G is a **cyclic group**;

if there is no such x, then G is **non-cyclic**.

18. Finite cyclic group

$\mathbb{Z}_6$ has order 6 and (two) elements of order 6,

so $\mathbb{Z}_6$ is cyclic;

$S(\square)$ has order 8 but no element of order 8,

so $S(\square)$ is non-cyclic.

Theorem 2.3

A finite group G of order n is cyclic

if and only if

G contains an element of order n.

19. Exercise 2.5

Find all the cyclic subgroups of

(a) $S(\triangle)$;   (b) $(\mathbb{Z}_5^*, \times_5)$;   (c) $(\mathbb{Z}_8, +_8)$.

Which of these groups are cyclic?

20. Infinite cyclic subgroup of G

Definition If $x \in G$ and there is **no** $n \in \mathbb{Z}^+$ such that $x^n = e$, then x has **infinite order**.

⟨x⟩ is the **infinite cyclic subgroup** generated by x.

G must be infinite

all powers of x are distinct

multiplicative $\langle x \rangle = \{\ldots, x^{-k}, \ldots, x^{-2}, x^{-1}, x^0 = e, x, x^2, \ldots, x^k, \ldots\}$

all powers

or

additive $\langle x \rangle = \{\ldots, -kx, \ldots, -2x, -x, 0x = 0, x, 2x, \ldots, kx, \ldots\}$.

all multiples

21. Examples

In $(\mathbb{R}^*, \times)$, $\langle 2 \rangle = \{\ldots, \frac{1}{4}, \frac{1}{2}, 1, 2, 4, \ldots\}$,

$\langle \frac{1}{2} \rangle = \{\ldots, 4, 2, 1, \frac{1}{2}, \frac{1}{4}, \ldots\} = \langle 2 \rangle$,

$\langle -2 \rangle = \{\ldots, \frac{1}{4}, -\frac{1}{2}, 1, -2, 4, \ldots\} = \langle -\frac{1}{2} \rangle$;

$(\mathbb{R}^*, \times)$ is non-cyclic

no element generates the whole of $\mathbb{R}^*$.

In $(\mathbb{Z}, +)$, $\langle 2 \rangle = \{\text{even integers}\} = 2\mathbb{Z} = \langle -2 \rangle$,

$\langle 3 \rangle = \{\text{multiples of } 3\} = 3\mathbb{Z} = \langle -3 \rangle$,

$\langle 1 \rangle = \{\ldots, -2, -1, 0, 1, 2, \ldots\} = \mathbb{Z} = \langle -1 \rangle$;

$\mathbb{Z}$ is cyclic

1 or -1 generates $\mathbb{Z}$.

22. Cyclic subgroups of S(O)

S(O) is infinite

$S(O) = \{r_\theta : \theta \in [0, 2\pi)\} \cup \{q_\phi : \phi \in [0, \pi)\}$

reflection q_ϕ : $\langle q_\phi \rangle = \{r_0, q_\phi\}$

order 2

rotation $r_{2\pi/5}$: $\langle r_{2\pi/5} \rangle = \{r_0, r_{2\pi/5}, r_{4\pi/5}, r_{6\pi/5}, r_{8\pi/5}\}$

order 5

$r_{2\pi/5}$

rotation r_1 : r_1 is rotation through 1 radian ($\simeq 57°$),

r_1^2 is rotation through 2 radians,

$\vdots$

r_1^7 is rotation through $7 - 2\pi$ radians,

$\vdots$

r_1^n is rotation through n radians (mod 2π).

π is irrational

infinite order

There is **no** $n \in \mathbb{Z}^+$ such that $r_1^n = r_0$,

so $\langle r_1 \rangle$ is an infinite cyclic subgroup.

23. Exercise 2.6

Find the order of each of the following elements in S(O):

$r_{\pi/4}$, $r_{\pi/3}$, $r_{2\pi/7}$, r_2.

2.2 Connections between cyclic groups

Having considered several examples of cyclic groups in the audio, we now examine some links between these groups. We start by considering two cyclic groups of order 4.

The group of rotations of the square, $S^+(\square)$, is a cyclic group that can be generated by a, the rotation through $\pi/2$. In Frame 7 we represented this diagrammatically as follows.

$$S^+(\square) = \{e = a^0, a^1, a^2, a^3\}$$

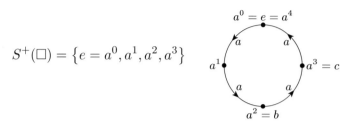

Compare this with the group $(\mathbb{Z}_4, +_4)$, which can be generated by the element 1. We can represent this group diagrammatically as follows.

$$\mathbb{Z}_4 = \{0, 1, 2, 3\}$$

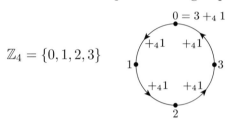

These two diagrammatic representations are very similar. The indices of the elements a^0, a^1, a^2, a^3 in the first set correspond exactly to the elements $0, 1, 2, 3$ in the second set. Also, on the first cycle diagram, moving around one place anticlockwise corresponds to composing with the rotation a. This results in adding 1 to the index which corresponds to moving around one place anticlockwise in the second cycle diagram.

There are similar connections between the direct symmetries of other regular polygons and the corresponding additive modular arithmetics:

$$
\begin{aligned}
S^+(\triangle) &= \{e = a^0, a^1, a^2\} & \text{and} \quad \mathbb{Z}_3 &= \{0, 1, 2\}, \\
S^+(\square) &= \{e = a^0, a^1, a^2, a^3\} & \text{and} \quad \mathbb{Z}_4 &= \{0, 1, 2, 3\}, \\
S^+(\pentagon) &= \{e = a^0, a^1, a^2, a^3, a^4\} & \text{and} \quad \mathbb{Z}_5 &= \{0, 1, 2, 3, 4\}, \\
S^+(\hexagon) &= \{e = a^0, a^1, a^2, a^3, a^4, a^5\} & \text{and} \quad \mathbb{Z}_6 &= \{0, 1, 2, 3, 4, 5\}, \\
&\ \ \vdots & &\ \ \vdots
\end{aligned}
$$

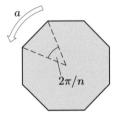

The generators of these groups of direct symmetries are the 'basic' rotations which move the polygon one place around as shown in the margin. For the regular n-gon, the symmetry group is $S^+(n\text{-GON}) = \langle a \rangle$, where a is a rotation through $2\pi/n$. Performing this rotation n times returns the polygon to its original position, so

$$S^+(n\text{-GON}) = \langle a \rangle = \{e, a, a^2, \ldots, a^{n-1}\}.$$

Note that $a^{n-1} = a^{-1}$, the inverse of a.

An alternative way to think of the members of $S^+(n\text{-GON})$ is 'move 0 places around' (the identity), 'move 1 place around', 'move 2 places around', ..., 'move $n-1$ places around'. In this way, the n-gon gives rise to a symmetry group with an additive structure—precisely the structure of $(\mathbb{Z}_n, +_n)$:

$$(\mathbb{Z}_n, +_n) = \langle 1 \rangle = \{0, 1, 2, \ldots, n-1\}.$$

In this sense, the group $(\mathbb{Z}_n, +_n)$ 'looks the same' as the group $S^+(n\text{-GON})$. We develop and formalise this idea of groups 'looking the same' in Section 3.

2.3 Proofs (optional)

In this section we give proofs of the two theorems that were stated without proof in the audio section. If you find the details hard to follow, do not be concerned, but do try to note the general strategies of the proofs. We start by proving the theorem stated in Frame 13.

These proofs are provided only for completeness. You do not need to be familiar with them.

Theorem 2.1   Let x be an element of a group G. If x has order n, then $\langle x \rangle$ has exactly n distinct elements:

$$\langle x \rangle = \left\{ e, x, x^2, \ldots, x^{n-2}, x^{n-1} \right\}.$$

The group G may be finite or infinite.

We give the proof in multiplicative notation.

Proof   Our proof is in two parts. In part (a) we show that the set

$$\left\{ e, x, x^2, \ldots, x^{n-2}, x^{n-1} \right\}$$

has exactly n elements, and in part (b) we show that the set is equal to $\langle x \rangle$.

(a) To show that the set has n elements we need to check that the elements $e, x, x^2, \ldots, x^{n-2}, x^{n-1}$ are distinct. The proof is by contradiction.

Suppose that

$$x^s = x^t, \quad \text{for some } s \text{ and } t \text{ with } 0 \le s < t \le n - 1.$$

Composing each side of this equation with x^{-s} on the right, we obtain

$$x^s x^{-s} = x^t x^{-s},$$

that is,

$$e = x^{t-s} \quad (\text{since } x^s x^{-s} = x^{s-s} = x^0 = e).$$

But $0 < t - s < n$, so this contradicts the definition of n as the order of x; that is, as the *least* positive integer such that $x^n = e$. Hence

$$x^s \ne x^t$$

and we conclude that $\left\{ e, x, x^2, \ldots, x^{n-2}, x^{n-1} \right\}$ has n elements.

(b) To show that $\left\{ e, x, x^2, \ldots, x^{n-2}, x^{n-1} \right\}$ is equal to $\langle x \rangle$, notice that the elements of $\left\{ e, x, x^2, \ldots, x^{n-2}, x^{n-1} \right\}$ belong to $\langle x \rangle$. So it is sufficient to check that the elements of $\langle x \rangle$ belong to $\left\{ e, x, x^2, \ldots, x^{n-2}, x^{n-1} \right\}$.

Each element of $\langle x \rangle$ has the form x^k, where $k \in \mathbb{Z}$. If we divide k by n we obtain a quotient q and a remainder r. We can therefore write

$$k = qn + r, \quad \text{for some } q, r \in \mathbb{Z}, \text{ where } 0 \le r \le n - 1.$$

Hence

$$\begin{aligned}
x^k &= x^{qn+r} \\
&= x^{qn} x^r \\
&= (x^n)^q x^r \\
&= e^q x^r \quad (\text{since } x^n = e) \\
&= x^r, \quad \text{where } 0 \le r \le n - 1.
\end{aligned}$$

So

$$x^k \in \left\{ e, x, x^2, \ldots, x^{n-2}, x^{n-1} \right\},$$

as required. ∎

> **Theorem 2.3**   Let G be a finite group of order n. Then G is cyclic if and only if G contains an element of order n.

See Frame 18.

Proof   First we prove the 'if' statement; that is,

IF G contains an element of order n, THEN G is cyclic.

Suppose that G contains an element x of order n.

By Theorem 2.1, the n elements

$$e, x, x^2, \ldots, x^{n-2}, x^{n-1}$$

are all distinct. But G has order n, so these elements constitute the whole group. Hence $\langle x \rangle = G$, so G is cyclic.

Next we prove the 'only if' statement; that is,

IF G is cyclic, THEN G contains an element of order n.

Suppose that G is a cyclic group of order n.

Then there is an element $x \in G$ which generates G:

$$G = \langle x \rangle = \left\{ x, x^2, \ldots, x^{n-2}, x^{n-1}, x^n = e \right\}.$$

It follows that n is the *least* positive integer such that $x^n = e$; that is, x has order n.   ∎

Further exercises

Exercise 2.7   Find the subgroup generated by each element of the following groups.

(a) $(\mathbb{Z}_{14}, +_{14})$ (b) $S(\triangle)$

Hence write down the order of each element in each of these groups.

Exercise 2.8   Decide which of the following groups are cyclic, and give all the generators of those that are cyclic.

(a) $(\{1, 2, 4, 7, 8, 11, 13, 14\}, \times_{15})$ (b) $(\{1, 3, 5, 7, 9, 11, 13, 15\}, \times_{16})$

(c) $(\mathbb{Z}_7^*, \times_7)$ (d) $(\mathbb{Z}_{11}^*, \times_{11})$ (e) $(\{1, 3, 5, 9, 11, 13\}, \times_{14})$

Exercise 2.9   Write down the order of each of the following elements of $S(\bigcirc)$.

(a) $r_{2\pi/9}$ (b) $r_{3\pi/7}$ (c) r_3

Exercise 2.10   Show that if $(G, \circ)$ is a cyclic group, then $(G, \circ)$ is Abelian.

3   Isomorphisms

After working through this section, you should be able to:

(a) arrange the elements in a group of order 4 or 6 so that the Cayley table exhibits one of two patterns;

(b) explain the meaning of the terms *isomorphic groups* and *isomorphism*;

(c) construct an isomorphism between two groups whose Cayley tables have the same pattern;

(d) recognise that any two cyclic groups of the same order are isomorphic;

(e) describe some standard properties of an isomorphism.

We ended the previous section by comparing the groups $(\mathbb{Z}_n, +_n)$ and $S(n\text{-GON}, \circ)$ of order n, and we saw that there is considerable similarity between them. In fact, these groups are 'essentially the same' in the sense that there is a one-one mapping from one onto the other which matches 'powers'. We call two groups that are 'essentially the same' *isomorphic groups*. In the video programme we look at some more examples of isomorphic groups, and then formalise the idea of isomorphism.

Watch the video programme 'Isomorphism'.

Video

3.1 Review of the programme

Cayley tables for groups of order 4

The programme begins with a Scottish dance sequence, from which we pick out the following four moves.

a Each couple moves around a quarter-turn anticlockwise.
b Each couple moves around a half-turn.
c Each couple moves around a quarter-turn clockwise.
e The four couples return to their original positions.

By considering the beginning and ending positions of these moves, we construct the following Cayley table for composing them.

$(D, \circ)$

$\circ$	e	a	b	c
e	e	a	b	c
a	a	b	c	e
b	b	c	e	a
c	c	e	a	b

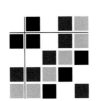

We observe that this table has a pattern of diagonal stripes in the body of the table. In fact, it is a group table; we call this group $(D, \circ)$.

Next, we look at some groups in modular arithmetic. The group $(\mathbb{Z}_4, +_4)$ has the following Cayley table, which has the same pattern of diagonal stripes.

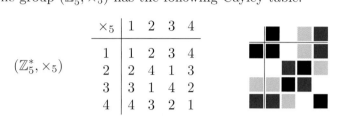

$(\mathbb{Z}_4, +_4)$

$+_4$	0	1	2	3
0	0	1	2	3
1	1	2	3	0
2	2	3	0	1
3	3	0	1	2

The two Cayley tables above exhibit identical patterns characterised by the diagonal stripes of elements. Do we know of any other groups of order 4 with tables which exhibit this pattern?

The group $(\mathbb{Z}_5^*, \times_5)$ has the following Cayley table.

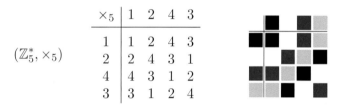

$(\mathbb{Z}_5^*, \times_5)$

$\times_5$	1	2	3	4
1	1	2	3	4
2	2	4	1	3
3	3	1	4	2
4	4	3	2	1

The pattern in this table appears to be different. However, we show that the table can be rearranged so that it has a pattern of diagonal stripes: we interchange the positions of the elements 3 and 4 in *both* borders of the table and then rearrange the entries in the body of the table accordingly, as shown below. (To rearrange the entries in the body of the table, we can either just fill them in again using the rearranged table borders, or we can take the original table, interchange the last two columns to obtain the intermediate table shown in the margin and then interchange the last two rows of this intermediate table to obtain the new table.)

We must rearrange both borders in the same way.

$\times_5$	1	2	4	3
1	1	2	4	3
2	2	4	3	1
3	3	1	2	4
4	4	3	1	2

$(\mathbb{Z}_5^*, \times_5)$

$\times_5$	1	2	4	3
1	1	2	4	3
2	2	4	3	1
4	4	3	1	2
3	3	1	2	4

It is important to remember that when we rearrange the entries in the borders of a Cayley table, we must rearrange *both* borders in the same way, since in a Cayley table the entries in the two borders must be in the same order.

So the groups $(\mathbb{Z}_4, +_4)$ and $(\mathbb{Z}_5^*, \times_5)$ both exhibit a pattern of diagonal stripes. Can every Cayley table of a group of order 4 be rearranged to exhibit this pattern?

To help us answer this question, we consider two other groups of order 4: the symmetry group of the rectangle and the set $\{1, 3, 5, 7\}$ under multiplication modulo 8.

$\circ$	e	a	r	s
e	e	a	r	s
a	a	e	s	r
r	r	s	e	a
s	s	r	a	e

$(S(\square), \circ)$

$\times_8$	1	3	5	7
1	1	3	5	7
3	3	1	7	5
5	5	7	1	3
7	7	5	3	1

$(\{1, 3, 5, 7\}, \times_8)$

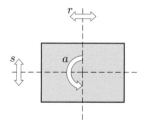

In Unit GTA1 we introduced $(S(\square), \circ)$ in Section 1 and the group $(\{1, 3, 5, 7\}, \times_8)$ in Section 3.

These tables exhibit a common pattern, but not the pattern with diagonal stripes. We observe that, in each of these groups, all the elements are self-inverse, so no rearrangement of the borders can remove any of the four identity elements from the leading diagonal of the table. So we cannot obtain the pattern of diagonal stripes for these two groups.

Thus we have found two different patterns for Cayley tables for groups of order 4:

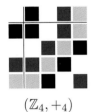

$(\mathbb{Z}_4, +_4)$

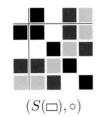

$(S(\square), \circ)$

For *any* group of order 4, we can rearrange, if necessary, the Cayley table so that it has one of the above two patterns: any group of order 4 has the same structure as one of the groups $(\mathbb{Z}_4, +_4)$ and $(S(\square), \circ)$.

We prove this in Unit GTA4.

Notice that both these patterns are symmetric about the leading diagonal, so any group of order 4 is Abelian.

Cayley tables for groups of order 6

Next, we consider groups of order 6. We consider the groups $(\mathbb{Z}_6, +_6)$ and $(S(\triangle), \circ)$, and examine their Cayley tables.

$(\mathbb{Z}_6, +_6)$

$+_6$	0	1	2	3	4	5
0	0	1	2	3	4	5
1	1	2	3	4	5	0
2	2	3	4	5	0	1
3	3	4	5	0	1	2
4	4	5	0	1	2	3
5	5	0	1	2	3	4

$(S(\triangle), \circ)$

$\circ$	e	a	b	r	s	t
e	e	a	b	r	s	t
a	a	b	e	t	r	s
b	b	e	a	s	t	r
r	r	s	t	e	a	b
s	s	t	r	b	e	a
t	t	r	s	a	b	e

We observe the diagonal stripes in the Cayley table for $(\mathbb{Z}_6, +_6)$, and we argue that no rearrangement of the borders in the table for $(S(\triangle), \circ)$ can give this pattern. This is because the identity element appears on the leading diagonal twice in the table for $(\mathbb{Z}_6, +_6)$, but four times in the table for $(S(\triangle), \circ)$.

For any positive integer n, the Cayley table for $(\mathbb{Z}_n, +_n)$ exhibits the pattern of diagonal stripes when we list the elements in the border in the natural order.

It follows that the group $(\mathbb{Z}_6, +_6)$ has two self-inverse elements, but $(S(\triangle), \circ)$ has four. Thus these two groups have *different* structures.

31

Thus we have found two different patterns for Cayley tables for groups of order 6:

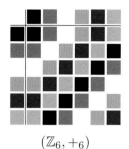   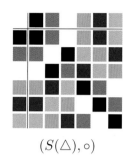

$(\mathbb{Z}_6, +_6)$ $(S(\triangle), \circ)$

For any group of order 6, we can rearrange, if necessary, the Cayley table so that it has one of these two patterns: any group of order 6 has the same structure as one of the groups $(\mathbb{Z}_6, +_6)$ and $(S(\triangle), \circ)$.

We prove this in Unit GTA4.

Notice that the pattern on the left is symmetric about the leading diagonal, and that a group with this pattern is an Abelian group; a group with the pattern on the right is non-Abelian.

Isomorphic groups

We say that two finite groups are 'essentially the same' if their Cayley tables can be rearranged so that they exhibit the same pattern. We wish to express this idea mathematically.

To help us formulate a definition, we look again at two groups of order 4.

$\circ$	e	a	b	c
e	e	a	b	c
a	a	b	c	e
b	b	c	e	a
c	c	e	a	b

$(D, \circ)$

$+_4$	0	1	2	3
0	0	1	2	3
1	1	2	3	0
2	2	3	0	1
3	3	0	1	2

$(\mathbb{Z}_4, +_4)$

To identify precisely what we mean when we say that the two groups are 'essentially the same', we try to find a mapping $\phi : D \longrightarrow \mathbb{Z}_4$ which is both one-one and onto; this ensures that we can match the borders of the Cayley tables. However, this is not sufficient—the entries in the bodies of the tables must also match correctly.

One-one and onto functions were introduced in Unit I2.

By comparing the tables, we see that the following one-one onto mapping does have the required property.

$$\phi : D \longrightarrow \mathbb{Z}_4$$
$$e \longmapsto 0$$
$$a \longmapsto 1$$
$$b \longmapsto 2$$
$$c \longmapsto 3$$

If we use this mapping to rename all the elements as they occur in the table for $(D, \circ)$, we get precisely the table for $(\mathbb{Z}_4, +_4)$.

$\circ$	e	a	b	c
e	e	a	b	c
a	a	b	c	e
b	b	c	e	a
c	c	e	a	b

$(D, \circ)$

	$\phi(e)$	$\phi(a)$	$\phi(b)$	$\phi(c)$
$\phi(e)$	$\phi(e)$	$\phi(a)$	$\phi(b)$	$\phi(c)$
$\phi(a)$	$\phi(a)$	$\phi(b)$	$\phi(c)$	$\phi(e)$
$\phi(b)$	$\phi(b)$	$\phi(c)$	$\phi(e)$	$\phi(a)$
$\phi(c)$	$\phi(c)$	$\phi(e)$	$\phi(a)$	$\phi(b)$

ϕ

$+_4$	0	1	2	3
0	0	1	2	3
1	1	2	3	0
2	2	3	0	1
3	3	0	1	2

$(\mathbb{Z}_4, +_4)$

The highlighted elements are referred to below.

For example, in $(D, \circ)$, we have

$a \circ b = c,$

and if we replace a by $\phi(a) = 1$, b by $\phi(b) = 2$, c by $\phi(c) = 3$ and $\circ$ by $+_4$, we obtain a corresponding entry in the table for $(\mathbb{Z}_4, +_4)$:

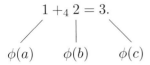

$1 +_4 2 = 3.$

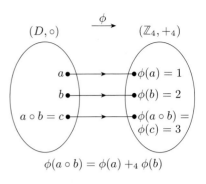

Thus we have

$\phi(a \circ b) = \phi(a) +_4 \phi(b),$

and all the other entries in the tables correspond in a similar way. In other words, the mapping ϕ has the following property.

> For each pair of elements in $(D, \circ)$, the image of their composite is the composite of their images in $(\mathbb{Z}_4, +_4)$.

Thus the two groups are 'essentially the same'.

Two groups which are 'essentially the same' are said to be *isomorphic*.

In general, if two groups $(G, \circ)$ and $(H, *)$ are isomorphic, then there must be a *one-one* mapping ϕ which maps G *onto* H such that

for all $g_1, g_2 \in G, \quad \phi(g_1 \circ g_2) = \phi(g_1) * \phi(g_2).$

For two general finite groups, we can picture the corresponding parts of the Cayley tables as follows.

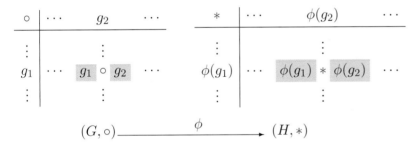

Now we can give a formal definition of *isomorphic groups*; the following definition applies both to finite and to infinite groups.

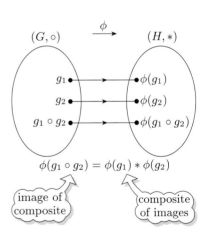

Definition   Two groups $(G, \circ)$ and $(H, *)$ are **isomorphic** if there exists a mapping $\phi : G \longrightarrow H$ such that both the following statements hold.

(a) ϕ is one-one and onto.

(b) For all $g_1, g_2 \in G$,

$\phi(g_1 \circ g_2) = \phi(g_1) * \phi(g_2).$

Such a function ϕ is called an **isomorphism**.

We write $(G, \circ) \cong (H, *)$ to denote that the groups $(G, \circ)$ and $(H, *)$ are isomorphic.

We sometimes abbreviate this to $G \cong H$, with the group operations being understood.

Remarks

1. An isomorphism is a mapping which is one-one and onto, so

    isomorphic groups $(G, \circ)$ and $(H, *)$ have the same order.

    Either $(G, \circ)$ and $(H, *)$ are both infinite groups, or $(G, \circ)$ and $(H, *)$ are both finite groups and $|G| = |H|$.

2. A family of groups can be divided into classes, called **isomorphism classes**, as follows. Two groups belong to the same isomorphism class if they are isomorphic, but to different classes otherwise; each group belongs to exactly one class.

    There are only two isomorphism classes for groups of order 4: the class of cyclic groups containing $(D, \circ)$, $(\mathbb{Z}_4, +_4)$, $(\mathbb{Z}_5^*, \times_5)$, ..., and the class containing $(S(\square), \circ)$, $(\{1, 3, 5, 7\}, \times_8)$, ... For convenience, we shall sometimes use the symbol C_4 to denote a typical group in the former class, and we shall refer to a typical group in the latter class as the **Klein group** and denote it by K_4. That is:

    C_4 denotes a typical cyclic group of order 4,

    K_4 denotes a typical group of order 4 in which each element is self-inverse.

    There are only two isomorphism classes for groups of order 6: the class containing $(\mathbb{Z}_6, +_6)$, ..., and the class containing $(S(\triangle), \circ)$, ...

3. Isomorphic groups possess a similar structure; for example:

    Any group isomorphic to an Abelian group is Abelian.

    Any group isomorphic to a cyclic group is cyclic.

    We prove these results in Unit GTB2.

Post-programme work

To show that two groups $(G, \circ)$ and $(H, *)$ are isomorphic, we have to find a function $\phi : (G, \circ) \longrightarrow (H, *)$ which satisfies the definition on page 33.

Strategy 3.1 To show that two groups $(G, \circ)$ and $(H, *)$ are isomorphic, show that there is a mapping $\phi : G \longrightarrow H$ such that:

1. ϕ is one-one and onto;

2. for all $g_1, g_2 \in G$,

    $$\phi(g_1 \circ g_2) = \phi(g_1) * \phi(g_2).$$

If $(G, \circ)$ and $(H, *)$ are finite groups of (the same) small order, it is sufficient to construct the Cayley tables for the two groups and to rearrange one of them to exhibit the same pattern as the other. Then write down a one-one onto mapping $\phi : G \longrightarrow H$ which matches up the Cayley tables.

If $(G, \circ)$ and $(H, *)$ are infinite groups or finite groups of (the same) large order, find a suitable mapping ϕ and show that it has properties 1 and 2.

The following example shows how Strategy 3.1 can be applied to show that two infinite groups are isomorphic. Recall that the infinite subgroup $\langle 2 \rangle$ of $(\mathbb{R}^*, \times)$ comprises the elements of the form 2^i, $i \in \mathbb{Z}$.

See Frames 2, 20 and 21.

Example 3.1   Show that $(\langle 2 \rangle, \times)$ is isomorphic to $(\mathbb{Z}, +)$ by showing that the following function ϕ is an isomorphism:

$$\phi \colon \langle 2 \rangle \longrightarrow \mathbb{Z}$$
$$2^i \longmapsto i.$$

Solution   We must show that the given mapping ϕ is one-one and onto, and also has the following property:

for all $2^i, 2^j \in \langle 2 \rangle$
$$\phi(2^i \times 2^j) = \phi(2^i) + \phi(2^j).$$

First we show that ϕ is one-one.

Suppose that $2^i, 2^j \in \langle 2 \rangle$ and that $\phi(2^i) = \phi(2^j)$; that is,

$$i = j.$$

Then $2^i = 2^j$ and so ϕ is one-one.

The mapping ϕ is onto because, for any element $i \in \mathbb{Z}$, there is a corresponding element $2^i \in \langle 2 \rangle$ such that $\phi(2^i) = i$.

Secondly, for all $2^i, 2^j \in \langle 2 \rangle$,

$$\phi(2^i \times 2^j) = \phi(2^{i+j}) = i + j = \phi(2^i) + \phi(2^j).$$

Hence ϕ is an isomorphism, so $(\langle 2 \rangle, \times) \cong (\mathbb{Z}, +)$.   ∎

Exercise 3.1   In each of the following cases, use Strategy 3.1 to show that the two groups given are isomorphic.

(a) $(\mathbb{Z}_4, +_4)$ and $(\mathbb{Z}_5^*, \times_5)$.

(b) $(D, \circ)$ and $(\mathbb{Z}_5^*, \times_5)$.

(c) $(S(\square), \circ)$ and $(\{1, 3, 5, 7\}, \times_8)$.

Exercise 3.2   Show that $(\mathbb{Z}, +)$ is isomorphic to $(6\mathbb{Z}, +)$ by showing that the following function ϕ is an isomorphism:

$$\phi \colon \mathbb{Z} \longrightarrow 6\mathbb{Z}$$
$$n \longmapsto 6n.$$

There is no general procedure for showing that two given groups are *not* isomorphic. Sometimes this may be clear—for example, if one group is cyclic or Abelian and the other is not, then they do not have the same structure.

We have the following partial strategy.

Strategy 3.2   To show that two *finite* groups $(G, \circ)$ and $(H, *)$ are *not* isomorphic, try any of the following methods.

- Compare the orders $|G|$ and $|H|$: if $|G| \neq |H|$, then $(G, \circ) \not\cong (H, *)$.

- Ascertain whether G and H are cyclic or Abelian: if one group is Abelian and the other is not, or if one group is cyclic and the other is not, then $(G, \circ) \not\cong (H, *)$.

- If the order is small, compare the entries in the leading diagonals of the group tables for G and H. For example, count the number of times the identity element appears and count the number of different elements that appear. If either of these counts differs between the two groups, then $(G, \circ) \not\cong (H, *)$.

Exercise 3.3   In each of the following cases, show that the two groups given are not isomorphic:

(a) $(\mathbb{Z}_8, +_8)$ and $(\mathbb{Z}_6, +_6)$;

(b) $(\mathbb{Z}_4, +_4)$ and $(\{1, 5, 7, 11\}, \times_{12})$.

3.2 Isomorphisms of cyclic groups

In Subsection 2.2 we showed that there are close connections between cyclic groups of the same order. In particular, we compared the cycle diagrams of the two groups $(S^+(\square), \circ)$ and $(\mathbb{Z}_4, +_4)$.

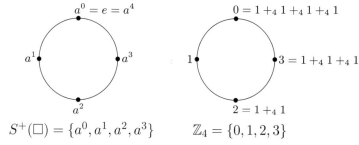

$$S^+(\square) = \{a^0, a^1, a^2, a^3\} \qquad \mathbb{Z}_4 = \{0, 1, 2, 3\}$$

The two cycle diagrams match exactly, and moving around one place anticlockwise in the first diagram (composing with a) corresponds to moving around one place anticlockwise in the second diagram (adding on 1, modulo 4).

We can now express these ideas more formally.

The cyclic groups $(S^+(\square), \circ)$ and $(\mathbb{Z}_4, +_4)$ are isomorphic, and the following mapping ϕ is an isomorphism:

$$\phi : S^+(\square) \longrightarrow \mathbb{Z}_4$$
$$a^1 \longmapsto 1$$
$$a^2 \longmapsto 2$$
$$a^3 \longmapsto 3$$
$$e = a^4 \longmapsto 0.$$

Notice how this isomorphism is obtained (see the following diagram).

$$\phi: \langle a \rangle \longrightarrow \langle 1 \rangle$$
$$a \longmapsto 1$$
$$a \circ a \longmapsto 1 +_4 1$$
$$a \circ a \circ a \longmapsto 1 +_4 1 +_4 1$$
$$a \circ a \circ a \circ a \longmapsto 1 +_4 1 +_4 1 +_4 1$$

we map a generator of $S^+(\square)$ to a generator of $\mathbb{Z}_4$

and then map 'powers' of a to like 'powers' of 1

This example illustrates the following theorem, which applies to groups of infinite order also.

Theorem 3.1   Two cyclic groups of the same order are isomorphic.

We shall prove this theorem in Unit GTB2.

There is a systematic way of finding isomorphisms between cyclic groups of the same order. The following strategy is a consequence of the proof of the theorem.

Strategy 3.3   To find an isomorphism between two finite cyclic groups G and H of the same order.

1.   Find a generator g of G and a generator h of H.

2.   Construct the following isomorphism ϕ:

$$\phi: G \longrightarrow H$$
$$g \longmapsto h$$
$$g^k \longmapsto h^k, \quad \text{for } k = 2, 3, \ldots.$$

$G = \langle g \rangle$ and $H = \langle h \rangle$.

Map g to h.

Map each 'power' of g to the corresponding 'power' of h.

Example 3.2   Find two isomorphisms between the groups $(\mathbb{Z}_4, +_4)$ and $(\mathbb{Z}_5^*, \times_5)$.

Solution   The group $(\mathbb{Z}_4, +_4)$ is generated by 1; the group $(\mathbb{Z}_5^*, \times_5)$ is generated by 2. Following Strategy 3.3, we obtain the following isomorphism from $(\mathbb{Z}_4, +_4)$ onto $(\mathbb{Z}_5^*, \times_5)$:

$$
\begin{aligned}
1 &\longmapsto 2 \\
1 +_4 1 &\longmapsto 2 \times_5 2 \\
1 +_4 1 +_4 1 &\longmapsto 2 \times_5 2 \times_5 2 \\
1 +_4 1 +_4 1 +_4 1 &\longmapsto 2 \times_5 2 \times_5 2 \times_5 2,
\end{aligned}
$$

that is,

$$
\begin{aligned}
1 &\longmapsto 2 \\
2 &\longmapsto 4 \\
3 &\longmapsto 3 \\
0 &\longmapsto 1.
\end{aligned}
$$

In Exercise 2.4(b), you found that the two elements 2 and 3 have order 4 in $(\mathbb{Z}_5^*, \times_5)$. They are both generators of this group.

This is the isomorphism obtained in the solution to Exercise 3.1(a).

However, $(\mathbb{Z}_4, +_4)$ is also generated by 3, so the following mapping is also an isomorphism:

$$
\begin{aligned}
3 &\longmapsto 2 \\
3 +_4 3 &\longmapsto 2 \times_5 2 \\
3 +_4 3 +_4 3 &\longmapsto 2 \times_5 2 \times_5 2 \\
3 +_4 3 +_4 3 +_4 3 &\longmapsto 2 \times_5 2 \times_5 2 \times_5 2,
\end{aligned}
$$

that is,

$$
\begin{aligned}
3 &\longmapsto 2 \\
2 &\longmapsto 4 \\
1 &\longmapsto 3 \\
0 &\longmapsto 1. \quad \blacksquare
\end{aligned}
$$

Theorem 3.1 tells us that any two cyclic groups of order n are isomorphic. Also, it is clear that we can construct a cyclic group of any order n by considering the rotations of a regular n-gon. If a is an anticlockwise rotation through $2\pi/n$, then $\langle a \rangle$ is a cyclic group with n elements—any other cyclic group of order n is isomorphic to $\langle a \rangle$. We shall find it convenient to introduce a symbol for a typical cyclic group of order n.

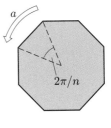

$\langle a \rangle = S^+(n\text{-GON})$

Notation   The symbol C_n denotes a typical cyclic group of order n, generated by x:

$$C_n = \langle x \rangle = \left\{ e, x, x^2, \ldots, x^{n-1} \right\}.$$

Exercise 3.4 Find all the generators of the cyclic group $(G, \times_9)$, where

$$G = \{1, 2, 4, 5, 7, 8\}.$$

Hence find two isomorphisms $\phi : (G, \times_9) \longrightarrow C_6$.

Exercise 3.5 Find all the generators of the cyclic group $(G, \times_{17})$, where

$$G = \{1, 2, 4, 8, 9, 13, 15, 16\}.$$

Hence find four isomorphisms $\phi : (G, \times_{17}) \longrightarrow C_8$.

3.3 Isomorphism classes of groups

The relation *is isomorphic to*, denoted by $\cong$, is an equivalence relation on the set of all groups, as we now show.

Equivalence relations were introduced in Unit I3.

E1 REFLEXIVE   Each group G is isomorphic to itself:

$$G \cong G.$$

We state these properties without proof.

E2 SYMMETRIC   If G and H are groups and G is isomorphic to H, then H is isomorphic to G:

$$\text{if } G \cong H, \quad \text{then} \quad H \cong G.$$

E3 TRANSITIVE   If G, H and K are groups and G is isomorphic to H and H is isomorphic to K, then G is isomorphic to K:

$$\text{if } G \cong H \quad \text{and} \quad H \cong K, \quad \text{then} \quad G \cong K.$$

We have seen that there are two isomorphism classes of groups of order 4. The class representatives are C_4, for the class of cyclic groups of order 4, and K_4, for the class of non-cyclic groups of order 4. Thus a group of order 4 is either isomorphic to C_4 or to K_4.

For example,

$$C_4 \cong S^+(\square) \cong S(\text{WIND}) \cong (\mathbb{Z}_4, +_4) \cong (\mathbb{Z}_5^*, \times_5)$$
$$\cong (\{0, 3, 6, 9\}, +_{12}) \cong (\{1, 9, 13, 17\}, \times_{20})$$
$$\cong (\{1, -1, i, -i\}, \times)$$

These lists are not exhaustive.

and

$$K_4 \cong S(\square) \cong (\{1, 3, 5, 7\}, \times_8) \cong (\{1, 5, 7, 11\}, \times_{12})$$
$$\cong (\{1, 4, 11, 14\}, \times_{15}) \cong (\{1, 7, 9, 15\}, \times_{16})$$
$$\cong (\{1, 9, 11, 19\}, \times_{20}).$$

Groups in the same isomorphism class possess a similar structure. For finite groups, their group tables exhibit the same patterns—once the elements have been reordered, if necessary.

3.4 Properties of isomorphisms

We conclude this section by listing some important properties of isomorphisms which were mentioned in the video programme.

We prove these properties in Unit GTB2.

Properties of isomorphisms

Let $(G, \circ)$ and $(H, *)$ be groups with identities e_G and e_H, respectively, and let $\phi : (G, \circ) \longrightarrow (H, *)$ be an isomorphism.

Then the isomorphism has the following properties.

1. IDENTITY ϕ matches the identity elements:
$$\phi(e_G) = e_H.$$

2. INVERSES ϕ matches inverses:
   for each $g \in G$,
$$\phi\big(g^{-1}\big) = (\phi(g))^{-1}.$$

3. POWERS ϕ matches 'powers':
   for each $g \in G$ and each $k \in \mathbb{Z}$,
$$\phi\big(g^k\big) = (\phi(g))^k.$$

For example, for the isomorphism
$$\phi : S^+(\square) \longrightarrow \mathbb{Z}_4 :$$
$$a^1 \longmapsto 1$$
$$a^2 \longmapsto 2$$
$$a^3 \longmapsto 3$$
$$e = a^4 \longmapsto 0.$$
$$\phi(e) = 0.$$

For example,
$$\phi\big(a^{-1}\big) = \phi\big(a^3\big) = 3$$
$$= 1^{-1} = (\phi(a))^{-1}.$$

For example,
$$\phi\big(a^2\big) = 2 = 1 +_4 1$$
$$= \phi(a) +_4 \phi(a).$$

We can illustrate these properties diagrammatically as follows.

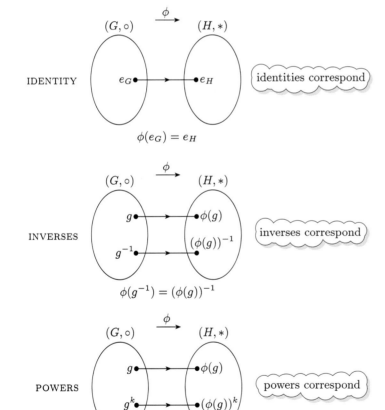

$$\phi(e_G) = e_H$$

$$\phi(g^{-1}) = (\phi(g))^{-1}$$

$$\phi(g^k) = (\phi(g))^k$$

Further exercises

Exercise 3.6 Find an isomorphism between the two groups in each of the following cases.

(a) $(\{1, 5, 7, 11\}, \times_{12})$ and $(\{1, 3, 5, 7\}, \times_8)$.

(b) $(\mathbb{Z}_{11}^*, \times_{11})$ and $(\mathbb{Z}_{10}, +_{10})$.

(c) $(\{1, 2, 4, 5, 7, 8\}, \times_9)$ and $(\{1, 3, 5, 9, 11, 13\}, \times_{14})$.

Exercise 3.7 Find the order of each element in each of the following groups.

(a) $(\{1, 2, 4, 7, 8, 11, 13, 14\}, \times_{15})$

(b) $(\{1, 3, 5, 7, 9, 11, 13, 15\}, \times_{16})$

Hint: Use the solutions to Exercise 2.8(a) and (b).

4   Groups from modular arithmetics

After working through this section, you should be able to:

(a) recognise that $(\mathbb{Z}_n, +_n)$ is a cyclic group;

(b) recognise that there is a cyclic subgroup $(\mathbb{Z}_m, +_n)$ of $(\mathbb{Z}_n, +_n)$ for each positive divisor m of n;

(c) identify all the generators of $(\mathbb{Z}_n, +_n)$;

(d) recognise that any subgroup of a cyclic group is cyclic also;

(e) recognise that $(\mathbb{Z}_p^*, \times_p)$ is a group, where p is a prime.

We have now seen a number of examples of groups in modular arithmetics. For example, in the video programme we considered the groups

$$(\mathbb{Z}_4, +_4), \quad (\mathbb{Z}_5^*, \times_5), \quad (\mathbb{Z}_6, +_6) \quad \text{and} \quad (\{1, 3, 5, 7\}, \times_8).$$

In this section we explore modular arithmetics further, and we discover many examples of groups and subgroups.

4.1 Additive modular arithmetics

We begin by asking you to do some exercises which illustrate some general results.

Exercise 4.1

(a) Construct the Cayley table for $(\mathbb{Z}_7, +_7)$ and use it to show that $(\mathbb{Z}_7, +_7)$ is a group.

(b) Verify that $(\mathbb{Z}_7, +_7)$ is cyclic by showing that 1 is a generator.

(c) Find all the cyclic subgroups of $(\mathbb{Z}_7, +_7)$ by finding the subgroup generated by each of its seven elements.

Exercise 4.2   Repeat the parts of Exercise 4.1 for the group $(\mathbb{Z}_{12}, +_{12})$.

You may find it helpful to refer to the solution to Exercise 2.5(c), where we considered $(\mathbb{Z}_8, +_8)$.

Exercises 4.1 and 4.2 illustrate some general results about additive modular arithmetics which we shall state, without proof, in this subsection. The most obvious is the following.

Theorem 4.1   For each $n \in \mathbb{N}$, $(\mathbb{Z}_n, +_n)$ is a cyclic group of order n and 1 is a generator.

Our second result concerns one of the differences between the group $(\mathbb{Z}_7, +_7)$ on the one hand and the groups $(\mathbb{Z}_8, +_8)$ and $(\mathbb{Z}_{12}, +_{12})$ on the other.

In Exercise 4.1 you found that $(\mathbb{Z}_7, +_7)$ is generated by each of its non-zero elements, but it is not true in general that $(\mathbb{Z}_n, +_n)$ is generated by each of its non-zero elements. The group $(\mathbb{Z}_8, +_8)$ has only four generators, 1, 3, 5 and 7, and the group $(\mathbb{Z}_{12}, +_{12})$ also has only four generators, 1, 5, 7 and 11. These generators are all primes—but before you jump to a conclusion about the generators of $(\mathbb{Z}_n, +_n)$, try the following exercise.

Exercise 4.3   Show that 9 generates the group $(\mathbb{Z}_{10}, +_{10})$.

In general, the group $(\mathbb{Z}_n, +_n)$ is generated by each element r that has no common factor with n, that is, is generated by each element r that is coprime to n.

For example, 9 is coprime to 10, but 5 is not. Coprime numbers are discussed in Unit I3.

Theorem 4.2   Let $r \in \mathbb{Z}_n$. Then r is a generator of $(\mathbb{Z}_n, +_n)$ if and only if r is coprime to n.

There is a further result, complementary to Theorem 4.2, that concerns the subgroups of $(\mathbb{Z}_n, +_n)$.

We have seen that $(\mathbb{Z}_8, +_8)$ has the following cyclic subgroups:

See Exercise 2.5(a).

$$\langle 0 \rangle = \{0\} \quad \text{of order 1,}$$
$$\langle 4 \rangle = \{0, 4\} \quad \text{of order 2,}$$
$$\langle 2 \rangle = \{0, 2, 4, 6\} \quad \text{of order 4,}$$
$$\langle 1 \rangle = \mathbb{Z}_8 \quad \text{of order 8.}$$

In Exercise 4.2 we found that $\mathbb{Z}_{12}$ has cyclic subgroups of orders 1, 2, 3, 4, 6 and 12. In each case, there is exactly one cyclic subgroup of $\mathbb{Z}_n$ for each divisor of n. This is our third result.

Theorem 4.3   The group $(\mathbb{Z}_n, +_n)$ has exactly one cyclic subgroup of order m for each divisor $m \in \mathbb{N}$ of n. This subgroup is either $\langle 0 \rangle$ (in the case where $m = 1$), or is generated by q, where $mq = n$.

For example, you found in Exercise 4.2 that $(\mathbb{Z}_{12}, +_{12})$ has exactly one cyclic subgroup of order 6—the subgroup $\langle 2 \rangle$.

The following result tells us that the group $(\mathbb{Z}_n, +_n)$ has no subgroups other than those given by Theorem 4.3.

Theorem 4.4   Let $(G, \circ)$ be a cyclic group. Then all the subgroups of $(G, \circ)$ are cyclic.

This result applies to infinite groups as well as finite ones.

In general, finding all the subgroups of a given finite group is an enormous task. Theorems 4.3 and 4.4 together give us a way of finding all the subgroups of a given finite *cyclic* group: we just find all its distinct cyclic subgroups.

Recall that every cyclic group of order n is isomorphic to $\mathbb{Z}_n$, so Theorems 4.3 and 4.4 tell us about all finite cyclic groups.

Exercise 4.4   Find all the subgroups of each of the following groups.

(a) $(\mathbb{Z}_9, +_9)$ (b) $(\mathbb{Z}_{10}, +_{10})$ (c) $(\mathbb{Z}_{11}, +_{11})$

4.2 Multiplicative modular arithmetics

In Subsection 4.1 we stated three theorems that describe the additive modular arithmetic groups. It is difficult to give similar results about multiplicative modular arithmetic groups, as the situation is much more complicated.

Exercise 4.5   Show that $(\mathbb{Z}_7, \times_7)$ is not a group.

We have met the kind of situation that occurs in Exercise 4.5 before. We found that $(\mathbb{R}, \times)$ fails to be a group because the number 0 has no multiplicative inverse in $\mathbb{R}$. Our solution in that situation was to discard 0 and consider the set $\mathbb{R}^* = \mathbb{R} - \{0\}$; we found that $(\mathbb{R}^*, \times)$ is a group. We saw also that $(\mathbb{Z}_5, \times_5)$ is not a group, but that if we discard 0 and consider $\mathbb{Z}_5^* = \{1, 2, 3, 4\}$, then $(\mathbb{Z}_5^*, \times_5)$ is a group of order 4. Here we generalise this approach and define

See Unit GTA1, Section 3, Frames 4 and 5.

See Unit GTA1, Exercise 3.5(b) and (c).

$$\mathbb{Z}_n^* = \mathbb{Z}_n - \{0\} = \{1, 2, 3, \ldots, n - 1\}.$$

Exercise 4.6

(a) Is $(\mathbb{Z}_6^*, \times_6)$ a group?

(b) Is $(\mathbb{Z}_9^*, \times_9)$ a group?

Justify your answer in each case.

The solution to Exercise 4.6 shows that we cannot always construct a group in multiplicative modular arithmetic simply by discarding 0 from our set of elements. The difficulty when we consider $\mathbb{Z}_6^*$ is that $2 \times_6 3 = 0 \notin \mathbb{Z}_6^*$, so the closure axiom fails. A similar result holds for $\mathbb{Z}_n^*$ whenever n is a product of two integers both greater than 1, say $n = rq$, because then we have $r \times_n q = 0$. However, if n is a *prime number*, then this situation cannot occur. We state the following general result without proof.

Theorem 4.5   Let p be a prime number. Then $(\mathbb{Z}_p^*, \times_p)$ is a group of order $p - 1$.

The proof of this theorem follows fairly easily from the corollary to Theorem 3.3, Unit I3.

Exercise 4.7   Show that $(\mathbb{Z}_{11}^*, \times_{11})$ is a cyclic group of order 10, and hence find all its subgroups.

Although $(\mathbb{Z}_n^*, \times_n)$ is not a group when n is not a prime, there do exist groups for which the binary operation is $\times_n$, where n is not a prime. For example, we have already met the groups $(\{1, 3, 5, 7\}, \times_8)$ and $(\{1, 4, 11, 14\}, \times_{15})$.

See Unit GTA1, Exercise 3.11(a and (c).

We state the following general result without proof.

> **Theorem 4.6**   For each $n \in \mathbb{N}$, the set of all numbers in $\mathbb{Z}_n$ that are coprime to n forms a group under $\times_n$.

The proof of this theorem follows fairly easily from the corollary to Theorem 3.3, Unit I3.

Other groups in multiplicative modular arithmetic are subgroups of the groups described in Theorem 4.6.

Exercise 4.8

(a) Show that $(\{1, 3, 7, 9, 11, 13, 17, 19\}, \times_{20})$ is a group.

(b) Find all the cyclic subgroups of $(\{1, 3, 7, 9, 11, 13, 17, 19\}, \times_{20})$, and hence show that this group is not cyclic.

(c) Show that

$$(\{1, 9, 11, 19\}, \times_{20}) \text{ is a subgroup of } (\{1, 3, 7, 9, 11, 13, 17, 19\}, \times_{20}),$$

but that it is not a cyclic subgroup.

Further exercises

Exercise 4.9   Find all the generators of each of the following groups.

(a) $(\mathbb{Z}_{14}, +_{14})$ (b) $(\mathbb{Z}_{16}, +_{16})$

Exercise 4.10   Find the order of each of the following elements in the given group.

(a) 3 in $(\mathbb{Z}_{15}, +_{15})$ (b) 9 in $(\mathbb{Z}_{21}, +_{21})$

Exercise 4.11   Find all the cyclic subgroups of the group

$$(\{1, 5, 7, 11, 13, 17, 19, 23\}, \times_{24}).$$

Exercise 4.12   Show that $(\mathbb{Z}_{13}^*, \times_{13})$ is a cyclic group, and hence find all its subgroups.

Solutions to the exercises

1.1 We have $\{e, b, s, u\} \subseteq S(\square)$, and the binary operation $\circ$ is the same on each set.

The Cayley table for $(\{e, b, s, u\}, \circ)$ is as follows.

$\circ$	e	b	s	u
e	e	b	s	u
b	b	e	u	s
s	s	u	e	b
u	u	s	b	e

We show that the three subgroup properties hold.

SG1   No new elements are needed to complete the table, so $\{e, b, s, u\}$ is closed under composition.

SG2   The identity in $S(\square)$ is e, and we have $e \in \{e, b, s, u\}$.

SG3   From the table, we see that each element of the set $\{e, b, s, u\}$ is self-inverse, so the set contains the inverse of each of its elements.

Hence $(\{e, b, s, u\}, \circ)$ satisfies the three subgroup properties, and so is a subgroup of $(S(\square), \circ)$.

1.2 (a) We have $3\mathbb{Z} \subseteq \mathbb{Z}$, and the binary operation $+$ is the same on each set.

We show that the three subgroup properties hold.

SG1   Let $x, y \in 3\mathbb{Z}$; then $x = 3m$ and $y = 3n$, for some $m, n \in \mathbb{Z}$. Thus
$$x + y = 3m + 3n = 3(m + n),$$
so $x + y \in 3\mathbb{Z}$.

Hence $3\mathbb{Z}$ is closed under addition.

SG2   The identity in $(\mathbb{Z}, +)$ is 0, and $0 = 3 \times 0 \in 3\mathbb{Z}$, so $3\mathbb{Z}$ contains the identity.

SG3   The inverse of $x = 3m$ is
$$-x = -3m = 3(-m) \in 3\mathbb{Z},$$
so $3\mathbb{Z}$ contains the inverse of each of its elements.

Hence $(3\mathbb{Z}, +)$ satisfies the three subgroup properties, and so is a subgroup of $(\mathbb{Z}, +)$.

(b) Here $6\mathbb{Z} \subseteq 2\mathbb{Z}$, and the binary operation $+$ is the same on each set.

We show that the three subgroup properties hold.

SG1   Let $x, y \in 6\mathbb{Z}$; then $x = 6m$, $y = 6n$, for some $m, n \in \mathbb{Z}$. So
$$x + y = 6m + 6n = 6(m + n) \in 6\mathbb{Z}.$$
Thus $6\mathbb{Z}$ is closed under $+$.

SG2   The identity in $(2\mathbb{Z}, +)$ is 0, and $0 = 6 \times 0$, so $0 \in 6\mathbb{Z}$.

Thus $6\mathbb{Z}$ contains the identity.

SG3   The inverse of $x = 6n$ is $-6n = 6(-n)$, and this belongs to $6\mathbb{Z}$.

Thus $6\mathbb{Z}$ contains the inverse of each of its elements.

Hence $(6\mathbb{Z}, +)$ satisfies the three subgroup properties, and so is a subgroup of $(2\mathbb{Z}, +)$.

1.3 The set $\mathbb{Q}^*$ contains negative numbers, but $\mathbb{R}^+$ does not. In particular, $-1 \in \mathbb{Q}^*$, but $-1 \notin \mathbb{R}^+$, so $\mathbb{Q}^* \nsubseteq \mathbb{R}^+$. It follows that $(\mathbb{Q}^*, \times)$ is not a subgroup of $(\mathbb{R}^+, \times)$.

1.4 In each case, we show that the three subgroup properties hold.

(a) The Cayley table for $(H_1, +_{12})$ is as follows.

$+_{12}$	0	3	6	9
0	0	3	6	9
3	3	6	9	0
6	6	9	0	3
9	9	0	3	6

SG1   No new elements are needed to complete the table, so H_1 is closed under $+_{12}$.

SG2   The identity in $(\mathbb{Z}_{12}, +_{12})$ is 0, and $0 \in H_1$.

SG3   From the table, we see that the inverse of each element of H_1 is in H_1.

Element	0	3	6	9
Inverse	0	9	6	3

Hence $(H_1, +_{12})$ satisfies the three subgroup properties, and so is a subgroup of $(\mathbb{Z}_{12}, +_{12})$.

(b) The Cayley table for $(H_2, +_{12})$ is as follows.

$+_{12}$	0	4	8
0	0	4	8
4	4	8	0
8	8	0	4

SG1   No new elements are needed to complete the table, so H_2 is closed under $+_{12}$.

SG2   The identity in $(\mathbb{Z}_{12}, +_{12})$ is 0, and $0 \in H_2$.

SG3   From the table, we see that the inverse of each element of H_2 is in H_2.

Element	0	4	8
Inverse	0	8	4

Hence $(H_2, +_{12})$ satisfies the three subgroup properties, and so is a subgroup of $(\mathbb{Z}_{12}, +_{12})$.

1.5 (a) Axiom SG1 fails: $a \circ a = b$, but $b \notin H$.

(b) Axiom SG1 fails: $2 \times_5 3 = 1$, but $1 \notin H$.
Alternatively, property SG2 fails: the identity in $\mathbb{Z}_5^*$ is 1, but $1 \notin H$.

(c) Axiom SG3 fails: for example, the inverse of 2 in $(\mathbb{R}^*, \times)$ is $\frac{1}{2}$, but $\frac{1}{2} \notin \mathbb{Z}^*$.

1.6 (a) We show that $(X, *)$ satisfies the four group axioms.

G1 Let $(a, b), (c, d) \in X$; then $a, b, c, d \in \mathbb{R}$, and $ab \neq 0$ and $cd \neq 0$, so $a \neq 0$, $b \neq 0$, $c \neq 0$ and $d \neq 0$.

   By definition,
   $$(a, b) * (c, d) = (ac, bd) \in \mathbb{R}^2;$$
   also, $(ac)(bd) = (ab)(cd) \neq 0$ because $ab \neq 0$ and $cd \neq 0$, so
   $$(a, b) * (c, d) \in X.$$
   So X is closed under $*$.

G2 Suppose that $(x, y) \in X$ is an identity in X. Then we must have, for each $(a, b) \in X$,
   $$(a, b) * (x, y) = (a, b);$$
   that is,
   $$(ax, by) = (a, b).$$
   Comparing coordinates, we obtain
   $$ax = a \quad \text{and} \quad by = b.$$
   Now $a \neq 0$ and $b \neq 0$, so we must have
   $$x = y = 1;$$
   so $(1, 1)$ acts as an identity on the right.
   If there is an identity, it must be $(1, 1)$.
   Also,
   $$(1, 1) * (a, b) = (a, b),$$
   so $(1, 1)$ acts as an identity on the left.
   The element $(1, 1) \in X$, since $1 \times 1 \neq 0$, and we have shown that, for each $(a, b) \in X$,
   $$(a, b) * (1, 1) = (a, b) = (1, 1) * (a, b),$$
   so $(1, 1)$ is an identity in X.

G3 Let $(a, b) \in X$; then $a, b \in \mathbb{R}$ and $ab \neq 0$, so $a \neq 0$ and $b \neq 0$.
   Suppose that $(a, b)^{-1} = (x, y)$, where $(x, y) \in X$. Then we must have
   $$(a, b) * (x, y) = (1, 1);$$
   that is,
   $$(ax, by) = (1, 1).$$
   Comparing coordinates, we obtain
   $$ax = 1 \quad \text{and} \quad by = 1.$$

Now $a \neq 0$ and $b \neq 0$, so we must have
$$x = 1/a \quad \text{and} \quad y = 1/b,$$
so $(1/a, 1/b)$ acts as an inverse on the right.
If (a, b) has an inverse, it must be $(1/a, 1/b)$.
Also,
$$(1/a, 1/b) * (a, b) = (1, 1),$$
so $(1/a, 1/b)$ acts as an inverse on the left.
Now $(1/a, 1/b) \in X$ because $(1/a, 1/b) \in \mathbb{R}^2$ and $(1/a)(1/b) = 1/ab \neq 0$.
We have shown that
$$(a, b) * (1/a, 1/b)$$
$$= (1, 1) = (1/a, 1/b) * (a, b),$$
so $(1/a, 1/b)$ is an inverse of (a, b).
Hence X contains an inverse of each of its elements.

G4 Let $(a, b), (c, d), (e, f) \in X$. We must show that
   $$(a, b) * ((c, d) * (e, f))$$
   $$= ((a, b) * (c, d)) * (e, f).$$
   First we consider the left-hand side:
   $$(a, b) * ((c, d) * (e, f))$$
   $$= (a, b) * (ce, df)$$
   $$= (a(ce), b(df))$$
   $$= (ace, bdf), \tag{S.1}$$
   since multiplication is associative on $\mathbb{R}$.
   Now we consider the right-hand side:
   $$((a, b) * (c, d)) * (e, f)$$
   $$= (ac, bd) * (e, f)$$
   $$= ((ac)e, (bd)f)$$
   $$= (ace, bdf), \tag{S.2}$$
   since multiplication is associative on $\mathbb{R}$.
   Expressions (S.1) and (S.2) are the same, so $*$ is associative on X.

Hence $(X, *)$ satisfies the four group axioms, and so is a group.

(b) $(A, *)$ is a subgroup of $(X, *)$.
We show that $(A, *)$ satisfies the three subgroup properties.

SG1 Let $(1, b), (1, d) \in A$; then $b \neq 0$ and $d \neq 0$.
   By definition,
   $$(1, b) * (1, d) = (1, bd);$$
   this element belongs to A because the first coordinate is 1 and the second coordinate is non-zero, since $b \neq 0$ and $d \neq 0$.
   So A is closed under $*$.

SG2 The identity in X is $(1,1)$, and $(1,1) \in A$ because the first coordinate is 1 and the second coordinate is non-zero.

So A contains the identity.

SG3 Let $(1,b) \in A$. Then $b \neq 0$ and
$$(1,b)^{-1} = (1, 1/b);$$
this element belongs to A because the first coordinate is 1 and the second coordinate is non-zero.

So A contains the inverse of each of its elements.

Hence $(A, *)$ satisfies the three subgroup properties, and so is a subgroup of $(X, *)$.

(c) $(B, *)$ is not a subgroup of $(X, *)$.

For example, $(3, -1)$ and $(4, -2)$ are in B, but
$$(3, -1) * (4, -2) = (12, 2) \notin B;$$
so B is not closed under $*$, and SG1 fails.

1.7

(a) (b) 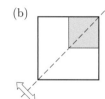 (c)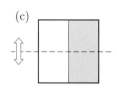

(a) The symmetry group is $\{e, a, b, c\} = S^+(\square)$.
(Any reflection interchanges the shaded and unshaded areas.)

(b) The symmetry group is $\{e, u\}$.
(The other elements of $S(\square)$ move the shaded square to a different corner.)

(c) The symmetry group is $\{e, t\}$.
(The other elements of $S(\square)$ move the shaded rectangle to other parts of the square.)

1.8

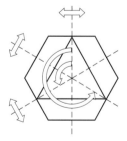

The elements of $S(F')$ are:

the identity,

rotations through $2\pi/3$ and $4\pi/3$ about the centre,

reflections in the three axes shown above.

The other elements of $S(\bigcirc)$ do not map the triangle to itself.

1.9 The required symbols are:
$$\begin{pmatrix} 1 & 2 & 3 & 4 \\ 1 & 2 & 3 & 4 \end{pmatrix}, \quad \begin{pmatrix} 1 & 2 & 3 & 4 \\ 2 & 4 & 3 & 1 \end{pmatrix}, \quad \begin{pmatrix} 1 & 2 & 3 & 4 \\ 4 & 1 & 3 & 2 \end{pmatrix},$$
$$\begin{pmatrix} 1 & 2 & 3 & 4 \\ 2 & 1 & 3 & 4 \end{pmatrix}, \quad \begin{pmatrix} 1 & 2 & 3 & 4 \\ 4 & 2 & 3 & 1 \end{pmatrix}, \quad \begin{pmatrix} 1 & 2 & 3 & 4 \\ 1 & 4 & 3 & 2 \end{pmatrix}.$$

Remark These two-line symbols are easily obtained by interchanging 3 and 4 in the two-line symbols for V_4, and then rearranging the columns so that the numbers in the top row are in the natural order.

1.10 (a) We describe the symmetries and we give the two-line symbol for each. You need to describe them in only one way. The symmetries of the modified figure are as follows.

The identity:
$$\begin{pmatrix} 1 & 2 & 3 & 4 & 5 & 6 \\ 1 & 2 & 3 & 4 & 5 & 6 \end{pmatrix},$$

a rotation through π about the vertical axis through the centre of the figure:
$$\begin{pmatrix} 1 & 2 & 3 & 4 & 5 & 6 \\ 4 & 6 & 5 & 1 & 3 & 2 \end{pmatrix},$$

reflection in the vertical plane through the locations 1 and 4:
$$\begin{pmatrix} 1 & 2 & 3 & 4 & 5 & 6 \\ 1 & 3 & 2 & 4 & 6 & 5 \end{pmatrix},$$

reflection in the vertical plane through the midpoints of the edges 14, 25 and 36:
$$\begin{pmatrix} 1 & 2 & 3 & 4 & 5 & 6 \\ 4 & 5 & 6 & 1 & 2 & 3 \end{pmatrix}.$$

Thus we obtain a subgroup of $S(F)$ of order 4.

(b) The symmetries of the modified figure are 'essentially the same' as those of $S(\triangle)$ as follows.

The identity:
$$\begin{pmatrix} 1 & 2 & 3 & 4 & 5 & 6 \\ 1 & 2 & 3 & 4 & 5 & 6 \end{pmatrix}.$$

Rotations through $2\pi/3$ and $4\pi/3$ about the horizontal axis of symmetry viewed from the near end:
$$\begin{pmatrix} 1 & 2 & 3 & 4 & 5 & 6 \\ 2 & 3 & 1 & 5 & 6 & 4 \end{pmatrix} \quad \text{and} \quad \begin{pmatrix} 1 & 2 & 3 & 4 & 5 & 6 \\ 3 & 1 & 2 & 6 & 4 & 5 \end{pmatrix}.$$

Reflections in the following planes of symmetry:

through the locations 1 and 4, and the midpoints of the edges 23 and 56
$$\begin{pmatrix} 1 & 2 & 3 & 4 & 5 & 6 \\ 1 & 3 & 2 & 4 & 6 & 5 \end{pmatrix},$$

through the locations 2 and 5, and the midpoints of the edges 13 and 46
$$\begin{pmatrix} 1 & 2 & 3 & 4 & 5 & 6 \\ 3 & 2 & 1 & 6 & 5 & 4 \end{pmatrix},$$

through the locations 3 and 6, and the midpoints of the edges 12 and 45

$$\begin{pmatrix} 1 & 2 & 3 & 4 & 5 & 6 \\ 2 & 1 & 3 & 5 & 4 & 6 \end{pmatrix}.$$

Thus we obtain a subgroup of $S(F)$ of order 6.

(c) The symmetries of the modified figure are as follows.

The identity:

$$\begin{pmatrix} 1 & 2 & 3 & 4 & 5 & 6 \\ 1 & 2 & 3 & 4 & 5 & 6 \end{pmatrix},$$

reflection in the plane through the locations 1 and 4, and the midpoints of the edges 23 and 56:

$$\begin{pmatrix} 1 & 2 & 3 & 4 & 5 & 6 \\ 1 & 3 & 2 & 4 & 6 & 5 \end{pmatrix}.$$

Thus we obtain a subgroup of $S(F)$ of order 2.

1.11 (a) Here $11\mathbb{Z} \subseteq \mathbb{Z}$, and the binary operation $+$ is the same on each set.

We show that the three subgroup properties hold.

SG1 Let $x, y \in 11\mathbb{Z}$; then $x = 11m$ and $y = 11n$, for some $m, n \in \mathbb{Z}$. Thus

$$x + y = 11m + 11n = 11(m + n),$$

and so $x + y \in 11\mathbb{Z}$.

Hence $11\mathbb{Z}$ is closed under addition.

SG2 The identity in $(\mathbb{Z}, +)$ is 0, and $0 = 11 \times 0 \in 11\mathbb{Z}$, so $11\mathbb{Z}$ contains the identity.

SG3 The inverse of $x = 11m$ is

$$-x = -11m = 11(-m) \in 11\mathbb{Z},$$

so $11\mathbb{Z}$ contains the inverse of each of its elements.

Hence $(11\mathbb{Z}, +)$ satisfies the three subgroup properties, and so is a subgroup of $(\mathbb{Z}, +)$.

(b) We have $\{10^k : k \in \mathbb{Z}\} \subseteq \mathbb{R}^*$, and the binary operation $\times$ is the same on each set.

We show that the three subgroup properties hold.

SG1 Let $x, y \in \{10^k : k \in \mathbb{Z}\}$; then $x = 10^m$ and $y = 10^n$, for some $m, n \in \mathbb{Z}$. Thus

$$xy = 10^m \times 10^n = 10^{m+n},$$

and so $xy \in \{10^k : k \in \mathbb{Z}\}$.

Hence $\{10^k : k \in \mathbb{Z}\}$ is closed under multiplication.

SG2 The identity in $(\mathbb{R}^*, \times)$ is 1, and

$$1 = 10^0 \in \{10^k : k \in \mathbb{Z}\},$$

so $\{10^k : k \in \mathbb{Z}\}$ contains the identity.

SG3 The inverse of $x = 10^m$ is

$$x^{-1} = 10^{-m} \in \{10^k : k \in \mathbb{Z}\},$$

so $\{10^k : k \in \mathbb{Z}\}$ contains the inverse of each of its elements.

Hence $(\{10^k : k \in \mathbb{Z}\}, \times)$ satisfies the three subgroup properties, and so is a subgroup of $(\mathbb{R}^*, \times)$.

(c) Here $\{z \in \mathbb{C} : z = x + ix\} \subseteq \mathbb{C}$, and the binary operation $+$ is the same on each set.

We show that the three subgroup properties hold.

SG1 Let $z_1 = x_1 + ix_1$ and $z_2 = x_2 + ix_2$; then

$$z_1 + z_2 = (x_1 + ix_1) + (x_2 + ix_2)$$
$$= (x_1 + x_2) + i(x_1 + x_2).$$

So $z_1 + z_2 \in \{z \in \mathbb{C} : z = x + ix\}$.

Hence $\{z \in \mathbb{C} : z = x + ix\}$ is closed under addition.

SG2 The identity in $(\mathbb{C}, +)$ is

$$0 = 0 + i0 \in \{z \in \mathbb{C} : z = x + ix\},$$

so $\{z \in \mathbb{C} : z = x + ix\}$ contains the identity.

SG3 The inverse of $z = v + iv$ is

$$-z = -(v + iv)$$
$$= -v + i(-v) \in \{z \in \mathbb{C} : z = x + ix\},$$

so $\{z \in \mathbb{C} : z = x + ix\}$ contains the inverse of each of its elements.

Hence $(\{z \in \mathbb{C} : z = x + ix\}, +)$ satisfies the three subgroup properties, and so is a subgroup of $(\mathbb{C}, +)$.

(d) We have $\{0, 3, 6, 9, 12\} \subseteq \mathbb{Z}_{15}$, and the binary operation $+_{15}$ is the same on each set.

We show that the three subgroup properties hold.

For this finite set, we use a Cayley table.

$+_{15}$	0	3	6	9	12
0	0	3	6	9	12
3	3	6	9	12	0
6	6	9	12	0	3
9	9	12	0	3	6
12	12	0	3	6	9

SG1 No new elements are needed to complete the table, so $\{0, 3, 6, 9, 12\}$ is closed under $+_{15}$.

SG2 The identity in $(\mathbb{Z}_{15}, +_{15})$ is 0, which is a member of the set.

SG3 From the table, we see that each element has an inverse in the set.

Element	0	3	6	9	12
Inverse	0	12	9	6	3

Hence $(\{0, 3, 6, 9, 12\}, +_{15})$ satisfies the three subgroup properties, and so is a subgroup of $(\mathbb{Z}_{15}, +_{15})$.

1.12 We have $C \subseteq X$, and the operation $*$ is the same on both sets. We show that $(C, *)$ satisfies the three subgroup properties.

SG1 Let $(a, 0), (c, 0) \in C$; then $a \neq 0$ and $c \neq 0$. By definition,
$$(a, 0) * (c, 0) = (ac, a \cdot 0 + 0) = (ac, 0);$$
this element belongs to C because the first coordinate is non-zero (since $a \neq 0$ and $c \neq 0$) and the second coordinate is zero.

So C is closed under $*$.

SG2 The identity in X is $(1, 0)$, and $(1, 0)$ belongs to C because the first coordinate is non-zero and the second coordinate is zero.

So C contains the identity.

SG3 Let $(a, 0) \in C$. Then $a \neq 0$, and
$$(a, 0)^{-1} = (1/a, -0/a) = (1/a, 0);$$
this element belongs to C because the first coordinate is non-zero and the second coordinate is zero.

So C contains the inverse of each of its elements.

Hence $(C, *)$ satisfies the three subgroup axioms, so $(C, *)$ is a subgroup of $(X, *)$.

1.13 **(a)** By the definition of $*$:
$$(-1, 3) * \left(\tfrac{1}{2}, 2\right) = \left(-1 \times 2 + 3 \times \tfrac{1}{2}, 3 \times 2\right) = \left(-\tfrac{1}{2}, 6\right),$$
$$(0, 3) * (-1, 4) = (0 \times 4 + 3 \times (-1), 3 \times 4) = (-3, 12).$$

(b) We have $A \subseteq X$, and the operation $*$ is the same on both sets. We show that $(A, *)$ satisfies the three subgroup properties.

SG1 Let $(0, b), (0, d) \in A$; then $b, d \in \mathbb{R}$, and $b \neq 0$ and $d \neq 0$.

By definition,
$$(0, b) * (0, d) = (0 \times d + b \times 0, bd) = (0, bd);$$
this element belongs to A because the first coordinate is 0 and the second coordinate is non-zero, since $b \neq 0$ and $d \neq 0$.

So A is closed under $*$.

SG2 The identity in X is $(0, 1)$, and $(0, 1) \in A$ because the first coordinate is 0 and the second coordinate is non-zero.

So A contains the identity.

SG3 Let $(0, b) \in A$. Then $b \neq 0$ and
$$(0, b)^{-1} = \left(0/b^2, 1/b\right) = (0, 1/b);$$
this element belongs to A because the first coordinate is 0 and the second coordinate is non-zero.

So A contains the inverse of each of its elements.

Hence $(A, *)$ satisfies the three subgroup properties and so is a subgroup of $(X, *)$.

(c) $B = \{(x, x) : x \in \mathbb{R}^*\}$. The identity in $(X, *)$ is $(0, 1)$, which does not belong to B.

Hence property SG2 fails, so $(B, *)$ is not a subgroup of $(X, *)$.

1.14 The symmetries are illustrated below.

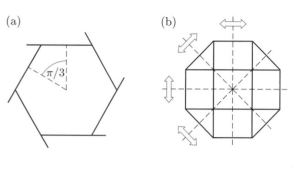

(a)

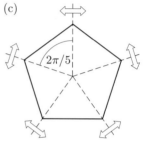

(b)

(c)

(a) The symmetries are the identity and rotations through $\pi/3$, $2\pi/3$, π, $4\pi/3$ and $5\pi/3$.

(b) The symmetries are the identity, rotations through $\pi/2$, π and $3\pi/2$, and reflections in the horizontal axis, the vertical axis and the two diagonal axes of symmetry.

(c) The figure has the same symmetries as the pentagon: the identity, rotations through multiples of $2\pi/5$ and five reflections.

1.15 The symmetries are illustrated below.

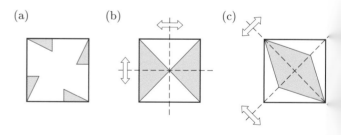

(a)   (b)   (c)

(a) The symmetries are the identity, and rotations through $\pi/2$, π and $3\pi/2$, so the group is $S^+(\square) = \{e, a, b, c\}$.

(b) The symmetries are the identity, rotation through π, and reflections in the horizontal and vertical axes of symmetry, so the group is $\{e, b, r, t\}$.

(c) The symmetries are the identity, rotation through π and reflections in the two diagonal axes of symmetry, so the group is $\{e, b, s, u\}$.

1.16

$$V_{12} = \left\{ e, \begin{pmatrix} 1 & 2 & 3 & 4 \\ 1 & 2 & 4 & 3 \end{pmatrix}, \begin{pmatrix} 1 & 2 & 3 & 4 \\ 2 & 1 & 3 & 4 \end{pmatrix}, \begin{pmatrix} 1 & 2 & 3 & 4 \\ 2 & 1 & 4 & 3 \end{pmatrix} \right\}.$$

2.1 (a)
$$\begin{aligned}
a \circ a &= b, \\
a \circ a \circ a &= b \circ a = c, \\
a \circ a \circ a \circ a &= c \circ a = e, \\
a \circ a \circ a \circ a \circ a &= e \circ a = a.
\end{aligned}$$

(b)
$$\begin{aligned}
a^{-1} \circ a^{-1} &= c \circ c = b, \\
a^{-1} \circ a^{-1} \circ a^{-1} &= b \circ c = a, \\
a^{-1} \circ a^{-1} \circ a^{-1} \circ a^{-1} &= a \circ c = e, \\
a^{-1} \circ a^{-1} \circ a^{-1} \circ a^{-1} \circ a^{-1} &= e \circ c = c = a^{-1}.
\end{aligned}$$

(c)
$$\begin{aligned}
b \circ b &= e, \\
b \circ b \circ b &= e \circ b = b.
\end{aligned}$$

(d)
$$\begin{aligned}
b^{-1} \circ b^{-1} &= b \circ b = e, \\
b^{-1} \circ b^{-1} \circ b^{-1} &= e \circ b = b = b^{-1}.
\end{aligned}$$

(e) Each of the reflections r, s, t and u is self-inverse, so composing each of these with itself gives the identity e:

$$r \circ r = s \circ s = t \circ t = u \circ u = e.$$

Subsequent composites alternate:

$$\begin{aligned}
r, e, r, e, r, \ldots & \quad \text{(for the reflection } r), \\
s, e, s, e, s, \ldots & \quad \text{(for the reflection } s), \\
t, e, t, e, t, \ldots & \quad \text{(for the reflection } t), \\
u, e, u, e, u, \ldots & \quad \text{(for the reflection } u).
\end{aligned}$$

2.2 (a)
$$\begin{aligned}
2 +_6 2 &= 4, \\
2 +_6 2 +_6 2 &= 4 +_6 2 = 0, \\
2 +_6 2 +_6 2 +_6 2 &= 0 +_6 2 = 2.
\end{aligned}$$

(b)
$$\begin{aligned}
4 +_6 4 &= 2, \\
4 +_6 4 +_6 4 &= 2 +_6 4 = 0, \\
4 +_6 4 +_6 4 +_6 4 &= 0 +_6 4 = 4.
\end{aligned}$$

(c)
$$\begin{aligned}
3 +_6 3 &= 0, \\
3 +_6 3 +_6 3 &= 0 +_6 3 = 3.
\end{aligned}$$

(d)
$$\begin{aligned}
1 +_6 1 &= 2, \\
1 +_6 1 +_6 1 &= 3, \\
1 +_6 1 +_6 1 +_6 1 &= 4, \\
1 +_6 1 +_6 1 +_6 1 +_6 1 &= 5.
\end{aligned}$$

(e)
$$\begin{aligned}
5 +_6 5 &= 4, \\
5 +_6 5 +_6 5 &= 4 +_6 5 = 3, \\
5 +_6 5 +_6 5 +_6 5 &= 3 +_6 5 = 2, \\
5 +_6 5 +_6 5 +_6 5 +_6 5 &= 2 +_6 5 = 1.
\end{aligned}$$

2.3 We use the results in Frame 8.

(a)

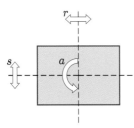

In $S(\square)$ each element is self-inverse, so

$$\langle e \rangle = \{e\}, \quad \langle a \rangle = \{e, a\},$$
$$\langle r \rangle = \{e, r\}, \quad \langle s \rangle = \{e, s\}.$$

(b) The Cayley table for $(\mathbb{Z}_5^*, \times_5)$ is as follows.

$\times_5$	1	2	3	4
1	1	2	3	4
2	2	4	1	3
3	3	1	4	2
4	4	3	2	1

The identity is 1, so

$$\langle 1 \rangle = \{1\}.$$

The powers of 2 are

$$\begin{aligned}
2 \times_5 2 &= 4, \\
2 \times_5 2 \times_5 2 &= 4 \times_5 2 = 3, \\
2 \times_5 2 \times_5 2 \times_5 2 &= 3 \times_5 2 = 1,
\end{aligned}$$

so

$$\langle 2 \rangle = \{1, 2, 4, 3\}.$$

The powers of 3 are

$$\begin{aligned}
3 \times_5 3 &= 4, \\
3 \times_5 3 \times_5 3 &= 4 \times_5 3 = 2, \\
3 \times_5 3 \times_5 3 \times_5 3 &= 2 \times_5 3 = 1,
\end{aligned}$$

so

$$\langle 3 \rangle = \{1, 3, 4, 2\}.$$

Notice that $3 = 2^{-1}$, so we could have written $\langle 3 \rangle = \langle 2^{-1} \rangle = \langle 2 \rangle$ directly.

The element 4 is self-inverse, so

$$\langle 4 \rangle = \{1, 4\}.$$

(c)

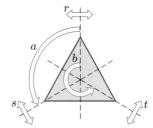

In $S(\triangle)$,

$$\langle e \rangle = \{e\}.$$

The powers of a are

$$\begin{aligned}
a \circ a &= b, \\
a \circ a \circ a &= b \circ a = e,
\end{aligned}$$

so

$$\langle a \rangle = \{e, a, b\}.$$

The powers of b are

$$\begin{aligned}
b \circ b &= a, \\
b \circ b \circ b &= a \circ b = e,
\end{aligned}$$

so

$$\langle b \rangle = \{e, b, a\}.$$

Alternatively, $\langle b \rangle = \langle a^{-1} \rangle = \langle a \rangle$.

The elements r, s and t are all self-inverse, so

$$\langle r \rangle = \{e, r\}, \quad \langle s \rangle = \{e, s\}, \quad \langle t \rangle = \{e, t\}.$$

2.4 We use the definition in Frame 10 and the solution to Exercise 2.3.

(a) For $S(\square)$, the orders are as follows.

Element	e	a	r	s
Order	1	2	2	2

(b) For $(\mathbb{Z}_5^*, \times_5)$, the orders are as follows.

Element	1	2	3	4
Order	1	4	4	2

(c) For $S(\triangle)$, the orders are as follows.

Element	e	a	b	r	s	t
Order	1	3	3	2	2	2

2.5 We use the definitions in Frames 15 and 17, and the solutions to Exercises 2.3 and 2.4.

(a) For $S(\triangle)$, the cyclic subgroups are

$$\{e\}, \quad \{e,a,b\}, \quad \{e,r\}, \quad \{e,s\}, \quad \{e,t\}.$$

No element generates the whole group, so $S(\triangle)$ is not cyclic.

Alternatively, $S(\triangle)$ has order 6 but no element of order 6, and so is not cyclic.

(b) For $(\mathbb{Z}_5^*, \times_5)$, the cyclic subgroups are

$$\{1\}, \quad \{1,4\}, \quad \{1,2,3,4\} = \mathbb{Z}_5^*,$$

so $(\mathbb{Z}_5^*, \times_5)$ is cyclic, with generators 2 and 3.

Alternatively, $(\mathbb{Z}_5^*, \times_5)$ has order 4 and (two) elements of order 4, and so is cyclic.

(c) For $(\mathbb{Z}_8, +_8)$, the cyclic subgroups are

$$\langle 0 \rangle = \{0\},$$
$$\langle 1 \rangle = \{0,1,2,3,4,5,6,7\} = \mathbb{Z}_8,$$
$$\langle 2 \rangle = \{0,2,4,6\},$$
$$\langle 3 \rangle = \{0,3,6,1,4,7,2,5\} = \mathbb{Z}_8,$$
$$\langle 4 \rangle = \{0,4\},$$
$$\langle 5 \rangle = \{0,5,2,7,4,1,6,3\} = \mathbb{Z}_8,$$
$$\langle 6 \rangle = \{0,6,4,2\},$$
$$\langle 7 \rangle = \{0,7,6,5,4,3,2,1\} = \mathbb{Z}_8.$$

So $(\mathbb{Z}_8, +_8)$ is cyclic, with generators 1, 3, 5 and 7.

Remark It is not necessary to calculate all the above sets independently. For example,

1 and 7 are inverses of each other,

so

$$\langle 1 \rangle = \langle -_8 1 \rangle = \langle 7 \rangle.$$

Alternatively, $(\mathbb{Z}_8, +_8)$ has order 8 and (four) elements of order 8, and so is cyclic.

2.6 $r_{\pi/4}$ has order 8:

$$\langle r_{\pi/4} \rangle = \{r_0, r_{\pi/4}, r_{\pi/2}, r_{3\pi/4}, r_\pi, r_{5\pi/4}, r_{3\pi/2}, r_{7\pi/4}\}.$$

$r_{\pi/3}$ has order 6:

$$\langle r_{\pi/3} \rangle = \{r_0, r_{\pi/3}, r_{2\pi/3}, r_\pi, r_{4\pi/3}, r_{5\pi/3}\}.$$

$r_{2\pi/7}$ has order 7:

$$\langle r_{2\pi/7} \rangle = \{r_0, r_{2\pi/7}, r_{4\pi/7}, r_{6\pi/7}, r_{8\pi/7}, r_{10\pi/7}, r_{12\pi/7}\}$$

r_2 has infinite order because

$$r_2^2 = r_4, \quad r_2^3 = r_6, \quad r_2^4 = r_8, \ldots ;$$

since π is irrational, no suffix is a multiple of 2π, so there is no positive integer n such that $r_2^n = r_0$ and so

$$\langle r \rangle = \{\ldots, r_{-6}, r_{-4}, r_{-2}, r_0, r_2, r_4, r_6, \ldots\}.$$

2.7 (a) The cyclic subgroups of $\mathbb{Z}_{14}$ are

$$\langle 0 \rangle = \{0\},$$
$$\langle 1 \rangle = \mathbb{Z}_{14} = \langle -_{14}1 \rangle = \langle 13 \rangle,$$
$$\langle 2 \rangle = \{0,2,4,6,8,10,12\} = \langle -_{14}2 \rangle = \langle 12 \rangle,$$
$$\langle 3 \rangle = \{0,3,6,9,12,1,4,7,10,13,2,5,8,11\}$$
$$= \mathbb{Z}_{14} = \langle -_{14}3 \rangle = \langle 11 \rangle,$$
$$\langle 4 \rangle = \{0,4,8,12,2,6,10\} = \langle -_{14}4 \rangle = \langle 10 \rangle,$$
$$\langle 5 \rangle = \{0,5,10,1,6,11,2,7,12,3,8,13,4,9\}$$
$$= \mathbb{Z}_{14} = \langle -_{14}5 \rangle = \langle 9 \rangle,$$
$$\langle 6 \rangle = \{0,6,12,4,10,2,8\} = \langle -_{14}6 \rangle = \langle 8 \rangle,$$
$$\langle 7 \rangle = \{0,7\}.$$

The order of each element in $\mathbb{Z}_{14}$ is the order of the subgroup it generates.

Element	0	1	2	3	4	5	6	7	8	9	10	11	12	13
Order	1	14	7	14	7	14	7	2	7	14	7	14	7	14

(b) In $S(\bigcirc)$:

$\langle e \rangle = \{e\}$, so e has order 1,

each rotation generates $S^+(\bigcirc)$, and so has order 5,

each reflection generates a subgroup of order 2 containing the identity and itself, and so has order 2.

2.8 In each group, we look at the subgroup generated by each element.

(a) The cyclic subgroups of $(\{1,2,4,7,8,11,13,14\}, \times_{15})$ are

$$\langle 1 \rangle = \{1\},$$
$$\langle 2 \rangle = \{1,2,4,8\} = \langle 2^{-1} \rangle = \langle 8 \rangle,$$
$$\langle 4 \rangle = \{1,4\},$$
$$\langle 7 \rangle = \{1,7,4,13\} = \langle 7^{-1} \rangle = \langle 13 \rangle,$$
$$\langle 11 \rangle = \{1,11\},$$
$$\langle 14 \rangle = \{1,14\}.$$

No element generates the whole group, so this group is not cyclic.

(b) The cyclic subgroups of $(\{1,3,5,7,9,11,13,15\}, \times_{16})$ are

$$\langle 1 \rangle = \{1\},$$
$$\langle 3 \rangle = \{1,3,9,11\} = \langle 3^{-1} \rangle = \langle 11 \rangle,$$
$$\langle 5 \rangle = \{1,5,9,13\} = \langle 5^{-1} \rangle = \langle 13 \rangle,$$
$$\langle 7 \rangle = \{1,7\},$$
$$\langle 9 \rangle = \{1,9\},$$
$$\langle 15 \rangle = \{1,15\}.$$

No element generates the whole group, so this group is not cyclic.

(c) The cyclic subgroups of $(\mathbb{Z}_7^*, \times_7)$ are

$\langle 1 \rangle = \{1\},$

$\langle 2 \rangle = \{1, 2, 4\} = \langle 2^{-1} \rangle = \langle 4 \rangle,$

$\langle 3 \rangle = \{1, 3, 2, 6, 4, 5\} = \mathbb{Z}_7^* = \langle 3^{-1} \rangle = \langle 5 \rangle,$

$\langle 6 \rangle = \{1, 6\}.$

This group is cyclic, with generators 3 and 5.

(d) The cyclic subgroups of $(\mathbb{Z}_{11}^*, \times_{11})$ are

$\langle 1 \rangle = \{1\},$

$\langle 2 \rangle = \{1, 2, 4, 8, 5, 10, 9, 7, 3, 6\}$
$= \mathbb{Z}_{11}^* = \langle 2^{-1} \rangle = \langle 6 \rangle,$

$\langle 3 \rangle = \{1, 3, 9, 5, 4\} = \langle 3^{-1} \rangle = \langle 4 \rangle,$

$\langle 5 \rangle = \{1, 5, 3, 4, 9\} = \langle 5^{-1} \rangle = \langle 9 \rangle,$

$\langle 7 \rangle = \{1, 7, 5, 2, 3, 10, 4, 6, 9, 8\}$
$= \mathbb{Z}_{11}^* = \langle 7^{-1} \rangle = \langle 8 \rangle,$

$\langle 10 \rangle = \{1, 10\}.$

This group is cyclic, with generators 2, 6, 7 and 8.

(e) The cyclic subgroups of $(\{1, 3, 5, 9, 11, 13\}, \times_{14})$ are

$\langle 1 \rangle = \{1\},$

$\langle 3 \rangle = \{1, 3, 9, 13, 11, 5\} = \langle 3^{-1} \rangle = \langle 5 \rangle,$

$\langle 9 \rangle = \{1, 9, 11\} = \langle 9^{-1} \rangle = \langle 11 \rangle,$

$\langle 13 \rangle = \{1, 13\}.$

This group is cyclic, with generators 3 and 5.

2.9 (a) $r_{2\pi/9}$ has order 9.

(b) $r_{3\pi/7}$ has order 14.

(c) r_3 has infinite order.

2.10 Let a be a generator of $(G, \circ)$. If g and h are elements of G, then we can write $g = a^n$ and $h = a^m$ for some $n, m \in \mathbb{Z}$, so

$g \circ h = a^n \circ a^m$
$= a^{n+m}$
$= a^{m+n}$
$= a^m \circ a^n$
$= h \circ g.$

Since g and h are arbitrary elements of G, it follows that $(G, \circ)$ is Abelian.

3.1 In each case, we see that the Cayley tables have the same pattern and we obtain an isomorphism by matching corresponding elements.

(a)

$+_4$	0	1	2	3
0	0	1	2	3
1	1	2	3	0
2	2	3	0	1
3	3	0	1	2

$(\mathbb{Z}_4, +_4)$

$\times_5$	1	2	4	3
1	1	2	4	3
2	2	4	3	1
4	4	3	1	2
3	3	1	2	4

$(\mathbb{Z}_5^*, \times_5)$

A suitable isomorphism is

$\phi : \mathbb{Z}_4 \longrightarrow \mathbb{Z}_5^*$

$0 \longmapsto 1$

$1 \longmapsto 2$

$2 \longmapsto 4$

$3 \longmapsto 3.$

(b)

$\circ$	e	a	b	c
e	e	a	b	c
a	a	b	c	e
b	b	c	e	a
c	c	e	a	b

$(D, \circ)$

$\times_5$	1	2	4	3
1	1	2	4	3
2	2	4	3	1
4	4	3	1	2
3	3	1	2	4

$(\mathbb{Z}_5^*, \times_5)$

A suitable isomorphism is

$\phi : D \longrightarrow \mathbb{Z}_5^*$

$e \longmapsto 1$

$a \longmapsto 2$

$b \longmapsto 4$

$c \longmapsto 3.$

(c)

$\circ$	e	a	r	s
e	e	a	r	s
a	a	e	s	r
r	r	s	e	a
s	s	r	a	e

$(S(\square), \circ)$

$\times_8$	1	3	5	7
1	1	3	5	7
3	3	1	7	5
5	5	7	1	3
7	7	5	3	1

$(\{1, 3, 5, 7\}, \times_8)$

A suitable isomorphism is

$\phi : S(\square) \longrightarrow \{1, 3, 5, 7\}$

$e \longmapsto 1$

$a \longmapsto 3$

$r \longmapsto 5$

$s \longmapsto 7.$

3.2 We must show that the given mapping ϕ is one-one and onto, and also has the following property:

for all $m, n \in \mathbb{Z}$,

$\phi(m + n) = \phi(m) + \phi(n).$

First we show that ϕ is one-one.

Suppose that $m, n \in \mathbb{Z}$ and that $\phi(m) = \phi(n)$; that is,

$6m = 6n.$

Then $m = n$ and so ϕ is one-one.

The mapping ϕ is onto because, for any element $6n \in 6\mathbb{Z}$, there is a corresponding element $n \in \mathbb{Z}$ such that $\phi(n) = 6n.$

Secondly, for all $m, n \in \mathbb{Z}$,

$\phi(m + n) = 6(m + n) = 6m + 6n = \phi(m) + \phi(n).$

Hence ϕ is an isomorphism, so $(\mathbb{Z}, +) \cong (6\mathbb{Z}, +).$

3.3 (a) $(\mathbb{Z}_8, +_8)$ has order 8, while $(\mathbb{Z}_6, +_6)$ has order 6, so no function $\mathbb{Z}_8 \longrightarrow \mathbb{Z}_6$ can be one-one and onto.

Hence these groups are not isomorphic.

(b) The group $(\mathbb{Z}_4 +_4)$ is cyclic, but $(\{1, 5, 7, 11\}, \times_{12})$ is not.

Hence these groups are not isomorphic.

Alternatively, construct the group tables and note that all the elements of $(\{1, 5, 7, 11\}, \times_{12})$ are self-inverse, so the leading diagonal of the Cayley table contains only the identity element. This is not the case for $(\mathbb{Z}_4, +_4)$ because the elements 1 and 3 are inverses of each other.

Hence these groups are not isomorphic.

3.4 The cyclic subgroups of $(G, \times_9)$, where $G = \{1, 2, 4, 5, 7, 8\}$, are

$\langle 1 \rangle = \{1\}$,

$\langle 2 \rangle = \{1, 2, 4, 8, 7, 5\}$,

$\langle 4 \rangle = \{1, 4, 7\}$,

$\langle 5 \rangle = \{1, 5, 7, 8, 4, 2\}$,

$\langle 7 \rangle = \{1, 7, 4\}$,

$\langle 8 \rangle = \{1, 8\}$.

So the elements 2 and 5 are generators of the group.

Remark It is not necessary to *calculate* the elements of $\langle 5 \rangle$ and $\langle 7 \rangle$ because

$\langle 2 \rangle = \langle 2^{-1} \rangle = \langle 5 \rangle$ and $\langle 4 \rangle = \langle 4^{-1} \rangle = \langle 7 \rangle$.

However, you may find it reassuring to use this property as a *check* (rather than as a shortcut).

The group C_6 is generated by x, where $x^6 = e$. Following Strategy 3.3 of mapping a generator to a generator, we obtain two possible isomorphisms:

$\phi_1 : G \longrightarrow C_6$	$\phi_2 : G \longrightarrow C_6$
$2^1 = 2 \longmapsto x$	$5^1 = 5 \longmapsto x$
$2^2 = 4 \longmapsto x^2$	$5^2 = 7 \longmapsto x^2$
$2^3 = 8 \longmapsto x^3$	$5^3 = 8 \longmapsto x^3$
$2^4 = 7 \longmapsto x^4$	$5^4 = 4 \longmapsto x^4$
$2^5 = 5 \longmapsto x^5$	$5^5 = 2 \longmapsto x^5$
$2^6 = 1 \longmapsto e$,	$5^6 = 1 \longmapsto e$.

which we can reorder as

$\phi_1 : G \longrightarrow C_6$	$\phi_2 : G \longrightarrow C_6$
$1 \longmapsto e$	$1 \longmapsto e$
$2 \longmapsto x$	$2 \longmapsto x^5$
$4 \longmapsto x^2$	$4 \longmapsto x^4$
$5 \longmapsto x^5$	$5 \longmapsto x$
$7 \longmapsto x^4$	$7 \longmapsto x^2$
$8 \longmapsto x^3$,	$8 \longmapsto x^3$.

Note that x^5 also generates C_6, but mapping the generators for G to x^5 results in the isomorphisms ϕ_1 and ϕ_2 above.

3.5 The cyclic subgroups of $(\{1, 2, 4, 8, 9, 13, 15, 16\}, \times_{17})$ are

$\langle 1 \rangle = \{1\}$,

$\langle 2 \rangle = \{1, 2, 4, 8, 16, 15, 13, 9\} = \langle 2^{-1} \rangle = \langle 9 \rangle$,

$\langle 4 \rangle = \{1, 4, 16, 13\} = \langle 4^{-1} \rangle = \langle 13 \rangle$,

$\langle 8 \rangle = \{1, 8, 13, 2, 16, 9, 4, 15\} = \langle 8^{-1} \rangle = \langle 15 \rangle$,

$\langle 16 \rangle = \{1, 16\}$.

So the elements 2, 8, 9 and 15 are generators of the group.

The group C_8 is generated by x, where $x^8 = e$. Following Strategy 3.3 of mapping a generator to a generator, we obtain four isomorphisms corresponding to $2 \longmapsto x$, $8 \longmapsto x$, $9 \longmapsto x$ and $15 \longmapsto x$, respectively.

$1 \longmapsto e$	$1 \longmapsto e$	$1 \longmapsto e$	$1 \longmapsto e$
$2 \longmapsto x$	$2 \longmapsto x^3$	$2 \longmapsto x^7$	$2 \longmapsto x^5$
$4 \longmapsto x^2$	$4 \longmapsto x^6$	$4 \longmapsto x^6$	$4 \longmapsto x^2$
$8 \longmapsto x^3$	$8 \longmapsto x$	$8 \longmapsto x^5$	$8 \longmapsto x^7$
$9 \longmapsto x^7$	$9 \longmapsto x^5$	$9 \longmapsto x$	$9 \longmapsto x^3$
$13 \longmapsto x^6$	$13 \longmapsto x^2$	$13 \longmapsto x^2$	$13 \longmapsto x^6$
$15 \longmapsto x^5$	$15 \longmapsto x^7$	$15 \longmapsto x^3$	$15 \longmapsto x$
$16 \longmapsto x^4$	$16 \longmapsto x^4$	$16 \longmapsto x^4$	$16 \longmapsto x^4$

3.6 (a) The Cayley tables are as follows.

$\times_{12}$	1	5	7	11
1	1	5	7	11
5	5	1	11	7
7	7	11	1	5
11	11	7	5	1

$\times_8$	1	3	5	7
1	1	3	5	7
3	3	1	7	5
5	5	7	1	3
7	7	5	3	1

The Cayley tables have the same pattern, so by matching corresponding elements, we obtain the following isomorphism:

$\phi : \{1, 5, 7, 11\} \longrightarrow \{1, 3, 5, 7\}$

$1 \longmapsto 1$

$5 \longmapsto 3$

$7 \longmapsto 5$

$11 \longmapsto 7$.

(b) By the solution to Exercise 2.8(d), $(\mathbb{Z}_{11}^*, \times_{11})$ is cyclic with generators 2, 6, 7 and 8. Also, $(\mathbb{Z}_{10}, +_{10})$ is cyclic and generated by 1. Using Strategy 3.3 to match powers of 2, 6, 7 and 8, gives any one of the following four isomorphisms, respectively.

$$\phi_1 : \mathbb{Z}_{11}^* \longrightarrow \mathbb{Z}_{10}$$
$$2^0 = 1 \longmapsto 0$$
$$2^1 = 2 \longmapsto 1$$
$$2^2 = 4 \longmapsto 2$$
$$2^3 = 8 \longmapsto 3$$
$$2^4 = 5 \longmapsto 4$$
$$2^5 = 10 \longmapsto 5$$
$$2^6 = 9 \longmapsto 6$$
$$2^7 = 7 \longmapsto 7$$
$$2^8 = 3 \longmapsto 8$$
$$2^9 = 6 \longmapsto 9$$

$$\phi_2 : \mathbb{Z}_{11}^* \longrightarrow \mathbb{Z}_{10}$$
$$6^0 = 1 \longmapsto 0$$
$$6^1 = 6 \longmapsto 1$$
$$6^2 = 3 \longmapsto 2$$
$$6^3 = 7 \longmapsto 3$$
$$6^4 = 9 \longmapsto 4$$
$$6^5 = 10 \longmapsto 5$$
$$6^6 = 5 \longmapsto 6$$
$$6^7 = 8 \longmapsto 7$$
$$6^8 = 4 \longmapsto 8$$
$$6^9 = 2 \longmapsto 9$$

$$\phi_3 : \mathbb{Z}_{11}^* \longrightarrow \mathbb{Z}_{10}$$
$$7^0 = 1 \longmapsto 0$$
$$7^1 = 7 \longmapsto 1$$
$$7^2 = 5 \longmapsto 2$$
$$7^3 = 2 \longmapsto 3$$
$$7^4 = 3 \longmapsto 4$$
$$7^5 = 10 \longmapsto 5$$
$$7^6 = 4 \longmapsto 6$$
$$7^7 = 6 \longmapsto 7$$
$$7^8 = 9 \longmapsto 8$$
$$7^9 = 8 \longmapsto 9$$

$$\phi_4 : \mathbb{Z}_{11}^* \longrightarrow \mathbb{Z}_{10}$$
$$8^0 = 1 \longmapsto 0$$
$$8^1 = 8 \longmapsto 1$$
$$8^2 = 9 \longmapsto 2$$
$$8^3 = 6 \longmapsto 3$$
$$8^4 = 4 \longmapsto 4$$
$$8^5 = 10 \longmapsto 5$$
$$8^6 = 3 \longmapsto 6$$
$$8^7 = 2 \longmapsto 7$$
$$8^8 = 5 \longmapsto 8$$
$$8^9 = 7 \longmapsto 9$$

(c) By the solution to Exercise 3.4, we know that the group $(G, \times_9)$, where $G = \{1, 2, 4, 5, 7, 8\}$, is cyclic with generators 2 and 5. Also, by the solution to Exercise 2.8(e), we know that the group $(H, \times_{14})$, where $H = \{1, 3, 5, 9, 11, 13\}$, is cyclic with generators 3 and 5.

Using Strategy 3.3 to match powers of the generators 2 and 5, gives either one of the following isomorphisms, respectively:

$$\phi_1 : G \longrightarrow H$$
$$1 \longmapsto 1$$
$$2 \longmapsto 3$$
$$2^2 \longmapsto 3^2$$
$$2^3 \longmapsto 3^3$$
$$2^4 \longmapsto 3^4$$
$$2^5 \longmapsto 3^5,$$

$$\phi_2 : G \longrightarrow H$$
$$1 \longmapsto 1$$
$$5 \longmapsto 3$$
$$5^2 \longmapsto 3^2$$
$$5^3 \longmapsto 3^3$$
$$5^4 \longmapsto 3^4$$
$$5^5 \longmapsto 3^5,$$

that is

$$\phi_1 : G \longrightarrow H$$
$$1 \longmapsto 1$$
$$2 \longmapsto 3$$
$$4 \longmapsto 9$$
$$8 \longmapsto 13$$
$$7 \longmapsto 11$$
$$5 \longmapsto 5,$$

$$\phi_2 : G \longrightarrow H$$
$$1 \longmapsto 1$$
$$5 \longmapsto 3$$
$$7 \longmapsto 9$$
$$8 \longmapsto 13$$
$$4 \longmapsto 11$$
$$2 \longmapsto 5.$$

3.7 (a) Using the solution to Exercise 2.8(a), we see that, in the group $(\{1, 2, 4, 7, 8, 11, 13, 14\}, \times_{15})$,

$\langle 1 \rangle = \{1\}$, so 1 has order 1,
$\langle 2 \rangle = \{1, 2, 4, 8\}$, so 2 has order 4,
$\langle 4 \rangle = \{1, 4\}$, so 4 has order 2,
$\langle 7 \rangle = \{1, 7, 4, 13\}$, so 7 has order 4,
$\langle 8 \rangle = \{1, 8, 4, 2\}$, so 8 has order 4,
$\langle 11 \rangle = \{1, 11\}$, so 11 has order 2,
$\langle 13 \rangle = \{1, 13, 4, 7\}$, so 13 has order 4,
$\langle 14 \rangle = \{1, 14\}$, so 14 has order 2.

(b) Using the solution to Exercise 2.8(b), we see that, in the group $(\{1, 3, 5, 7, 9, 11, 13, 15\}, \times_{16})$,

$\langle 1 \rangle = \{1\}$, so 1 has order 1,
$\langle 3 \rangle = \{1, 3, 9, 11\}$, so 3 has order 4,
$\langle 5 \rangle = \{1, 5, 9, 13\}$, so 5 has order 4,
$\langle 7 \rangle = \{1, 7\}$, so 7 has order 2,
$\langle 9 \rangle = \{1, 9\}$, so 9 has order 2,
$\langle 11 \rangle = \{1, 11, 9, 3\}$, so 11 has order 4,
$\langle 13 \rangle = \{1, 13, 9, 5\}$, so 13 has order 4,
$\langle 15 \rangle = \{1, 15\}$, so 15 has order 2.

4.1 (a) The Cayley table for $(\mathbb{Z}_7, +_7)$ is as follows.

$+_7$	0	1	2	3	4	5	6
0	0	1	2	3	4	5	6
1	1	2	3	4	5	6	0
2	2	3	4	5	6	0	1
3	3	4	5	6	0	1	2
4	4	5	6	0	1	2	3
5	5	6	0	1	2	3	4
6	6	0	1	2	3	4	5

We show that the four group axioms hold.

G1 No new elements are needed to complete the table, so $\mathbb{Z}_7$ is closed under $+_7$.

G2 From the table, we see that 0 is an identity element.

G3 From the table, we see that each element in $\mathbb{Z}_7$ has an inverse in $\mathbb{Z}_7$.

Element	0	1	2	3	4	5	6
Inverse	0	6	5	4	3	2	1

G4 Addition modulo 7 is associative.

Hence $(\mathbb{Z}_7, +_7)$ satisfies the four group axioms, and so is a group.

(b) We have

$$1 +_7 1 = 2, \quad 2 +_7 1 = 3, \quad 3 +_7 1 = 4,$$
$$4 +_7 1 = 5, \quad 5 +_7 1 = 6, \quad 6 +_7 1 = 0,$$

so

$$\langle 1 \rangle = \{1, 2, 3, 4, 5, 6, 0\} = \mathbb{Z}_7.$$

Thus $(\mathbb{Z}_7, +_7)$ is cyclic and 1 is a generator.

(c) The cyclic subgroups of the group $(\mathbb{Z}_7, +_7)$ are:

$$\langle 0 \rangle = \{0\},$$
$$\langle 1 \rangle = \mathbb{Z}_7 = \langle -_7 1 \rangle = \langle 6 \rangle,$$
$$\langle 2 \rangle = \{0, 2, 4, 6, 1, 3, 5\} = \mathbb{Z}_7 = \langle -_7 2 \rangle = \langle 5 \rangle,$$
$$\langle 3 \rangle = \{0, 3, 6, 2, 5, 1, 4\} = \mathbb{Z}_7 = \langle -_7 3 \rangle = \langle 4 \rangle.$$

Thus 2, 3, 4, 5 and 6 are also generators of $(\mathbb{Z}_7, +_7)$, and the only proper subgroup is the trivial subgroup.

4.2 (a) The Cayley table for $(\mathbb{Z}_{12}, +_{12})$ is as follows.

$+_{12}$	0	1	2	3	4	5	6	7	8	9	10	11
0	0	1	2	3	4	5	6	7	8	9	10	11
1	1	2	3	4	5	6	7	8	9	10	11	0
2	2	3	4	5	6	7	8	9	10	11	0	1
3	3	4	5	6	7	8	9	10	11	0	1	2
4	4	5	6	7	8	9	10	11	0	1	2	3
5	5	6	7	8	9	10	11	0	1	2	3	4
6	6	7	8	9	10	11	0	1	2	3	4	5
7	7	8	9	10	11	0	1	2	3	4	5	6
8	8	9	10	11	0	1	2	3	4	5	6	7
9	9	10	11	0	1	2	3	4	5	6	7	8
10	10	11	0	1	2	3	4	5	6	7	8	9
11	11	0	1	2	3	4	5	6	7	8	9	10

We show that the four group axioms hold.

G1   No new elements are needed to complete the table, so $\mathbb{Z}_{12}$ is closed under $+_{12}$.

G2   From the table, we see that 0 is an identity element.

G3   From the table, we see that each element in $\mathbb{Z}_{12}$ has an inverse in $\mathbb{Z}_{12}$.

Element	0	1	2	3	4	5	6	7	8	9	10	11
Inverse	0	11	10	9	8	7	6	5	4	3	2	1

G4   Addition modulo 12 is associative.

Hence $(\mathbb{Z}_{12}, +_{12})$ satisfies the four group axioms, and so is a group.

(b) We have
$$1 +_{12} 1 = 2, \quad 2 +_{12} 1 = 3, \quad 3 +_{12} 1 = 4,$$
$$4 +_{12} 1 = 5, \quad 5 +_{12} 1 = 6, \quad 6 +_{12} 1 = 7,$$
$$7 +_{12} 1 = 8, \quad 8 +_{12} 1 = 9, \quad 9 +_{12} 1 = 10,$$
$$10 +_{12} 1 = 11, \quad 11 +_{12} 1 = 0,$$
so
$$\langle 1 \rangle = \{1, 2, 3, 4, 5, 6, 7, 8, 9, 10, 11, 0\} = \mathbb{Z}_{12}.$$
Thus $(\mathbb{Z}_{12}, +_{12})$ is cyclic and 1 is a generator.

(c) The cyclic subgroups of the group $(\mathbb{Z}_{12}, +_{12})$ are:
$$\langle 0 \rangle = \{0\},$$
$$\langle 1 \rangle = \mathbb{Z}_{12} = \langle -_{12} 1 \rangle = \langle 11 \rangle,$$
$$\langle 2 \rangle = \{0, 2, 4, 6, 8, 10\} = \langle -_{12} 2 \rangle = \langle 10 \rangle,$$
$$\langle 3 \rangle = \{0, 3, 6, 9\} = \langle -_{12} 3 \rangle = \langle 9 \rangle,$$
$$\langle 4 \rangle = \{0, 4, 8\} = \langle -_{12} 4 \rangle = \langle 8 \rangle,$$
$$\langle 5 \rangle = \{0, 5, 10, 3, 8, 1, 6, 11, 4, 9, 2, 7\}$$
$$\quad = \mathbb{Z}_{12} = \langle -_{12} 5 \rangle = \langle 7 \rangle,$$
$$\langle 6 \rangle = \{0, 6\}.$$
Thus 5, 7 and 11 are also generators of $(\mathbb{Z}_{12}, +_{12})$.

4.3 Adding 9 to itself repeatedly, we obtain
$$9 +_{10} 9 = 8, \quad 8 +_{10} 9 = 7, \quad 7 +_{10} 9 = 6,$$
$$6 +_{10} 9 = 5, \quad 5 +_{10} 9 = 4, \quad 4 +_{10} 9 = 3,$$
$$3 +_{10} 9 = 2, \quad 2 +_{10} 9 = 1, \quad 1 +_{10} 9 = 0,$$
so
$$\langle 9 \rangle = \{0, 1, 2, 3, 4, 5, 6, 7, 8, 9\} = \mathbb{Z}_{10}.$$

4.4 The given groups are all cyclic, so, by Theorem 4.4, every subgroup is cyclic.

(a) By Theorem 4.3, there are three subgroups, with orders 1, 3 and 9 (the divisors of 9), and they are given by
$$\langle 0 \rangle = \{0\},$$
$$\langle 3 \rangle = \{0, 3, 6\},$$
$$\langle 1 \rangle = \{0, 1, 2, 3, 4, 5, 6, 7, 8\} = \mathbb{Z}_9.$$

(b) By Theorem 4.3, there are four subgroups, with orders 1, 2, 5 and 10 (the divisors of 10), and they are given by
$$\langle 0 \rangle = \{0\},$$
$$\langle 5 \rangle = \{0, 5\},$$
$$\langle 2 \rangle = \{0, 2, 4, 6, 8\},$$
$$\langle 1 \rangle = \{0, 1, 2, 3, 4, 5, 6, 7, 8, 9\} = \mathbb{Z}_{10}.$$

(c) Each non-zero element in $(\mathbb{Z}_{11}, +_{11})$ is coprime to 11, and therefore generates the whole group, by Theorem 4.2. So $(\mathbb{Z}_{11}, +_{11})$ has no subgroups other than itself and $\langle 0 \rangle$.

4.5 The identity in $(\mathbb{Z}_7, \times_7)$ is 1. Thus, to find an inverse of an element a in $\mathbb{Z}_7$, we need to solve the equation
$$a \times_7 x = 1.$$
However, for all $x \in \mathbb{Z}_7$,
$$0 \times_7 x = 0,$$
so 0 does not have a multiplicative inverse, and axiom G3 fails.

Hence $(\mathbb{Z}_7, \times_7)$ is not a group.

4.6 (a) $(\mathbb{Z}_6^*, \times_6)$ is not a group.

For example, $2, 3 \in \mathbb{Z}_6^*$, but $2 \times_6 3 = 0 \notin \mathbb{Z}_6^*$, so the closure axiom G1 fails.

(b) $(\mathbb{Z}_9^*, \times_9)$ is not a group.

For example, $3 \in \mathbb{Z}_9^*$, but $3 \times_9 3 = 0 \notin \mathbb{Z}_9^*$, so the closure axiom G1 fails.

4.7 We find the cyclic subgroup generated by each element of $(\mathbb{Z}_{11}^*, \times_{11})$:

$$\langle 1 \rangle = \{1\},$$
$$\langle 2 \rangle = \{1, 2, 4, 8, 5, 10, 9, 7, 3, 6\}$$
$$= \mathbb{Z}_{11}^* = \langle 2^{-1} \rangle = \langle 6 \rangle,$$
$$\langle 3 \rangle = \{1, 3, 9, 5, 4\} = \langle 3^{-1} \rangle = \langle 4 \rangle,$$
$$\langle 5 \rangle = \{1, 5, 3, 4, 9\} = \langle 5^{-1} \rangle = \langle 9 \rangle,$$
$$\langle 7 \rangle = \{1, 7, 5, 2, 3, 10, 4, 6, 9, 8\}$$
$$= \mathbb{Z}_{11}^* = \langle 7^{-1} \rangle = \langle 8 \rangle,$$
$$\langle 10 \rangle = \{1, 10\}.$$

Hence $(\mathbb{Z}_{11}^*, \times_{11})$ is a cyclic group with generators 2, 6, 7 and 8.

By Theorem 4.4, all the subgroups are cyclic, so the subgroups are those found above:

$$\{1\}, \quad \{1, 10\}, \quad \{1, 3, 4, 5, 9\} \quad \text{and} \quad \mathbb{Z}_{11}^* \text{ itself.}$$

4.8 (a) The set of all numbers coprime to 20 is $\{1, 3, 7, 9, 11, 13, 17, 19\}$, so, by Theorem 4.6, $(\{1, 3, 7, 9, 11, 13, 17, 19\}, \times_{20})$ is a group.

(b) The cyclic subgroups of this group are

$$\langle 1 \rangle = \{1\},$$
$$\langle 3 \rangle = \{1, 3, 9, 7\} = \langle 3^{-1} \rangle = \langle 7 \rangle,$$
$$\langle 9 \rangle = \{1, 9\},$$
$$\langle 11 \rangle = \{1, 11\},$$
$$\langle 13 \rangle = \{1, 13, 9, 17\} = \langle 13^{-1} \rangle = \langle 17 \rangle,$$
$$\langle 19 \rangle = \{1, 19\}.$$

No element generates the whole group, so the group $(\{1, 3, 7, 9, 11, 13, 17, 19\}, \times_{20})$ is not cyclic.

(c) The Cayley table for $(\{1, 9, 11, 19\}, \times_{20})$ is as follows.

$\times_{20}$	1	9	11	19
1	1	9	11	19
9	9	1	19	11
11	11	19	1	9
19	19	11	9	1

We show that the three subgroup properties hold.

SG1 No new elements are needed to complete the table, so $\{1, 9, 11, 19\}$ is closed under $\times_{20}$.

SG2 The identity element of the group is 1, and $1 \in \{1, 9, 11, 19\}$.

SG3 Each element in $\{1, 9, 11, 19\}$ is self-inverse, so the set contains the inverse of each of its elements.

Hence $(\{1, 9, 11, 19\}, \times_{20})$ satisfies the three subgroup properties, and so is a subgroup.

Finally, $(\{1, 9, 11, 19\}, \times_{20})$ is not cyclic: each element is self-inverse, so no element generates the whole subgroup $\{1, 9, 11, 19\}$.

4.9 (a) By Theorem 4.2, the generators of $(\mathbb{Z}_{14}, +_{14})$ are those integers which are coprime to 14: namely, 1, 3, 5, 9, 11 and 13. (See also the solution to Exercise 2.7(a).)

(b) By Theorem 4.2, the generators of $(\mathbb{Z}_{16}, +_{16})$ are those integers which are coprime to 16: namely, 1, 3, 5, 7, 9, 11, 13 and 15.

4.10 (a) In $(\mathbb{Z}_{15}, +_{15})$,

$$\langle 3 \rangle = \{0, 3, 6, 9, 12\},$$

so 3 has order 5.

(b) In $(\mathbb{Z}_{21}, +_{21})$,

$$\langle 9 \rangle = \{0, 9, 18, 6, 15, 3, 12\},$$

so 9 has order 7.

4.11 We find the cyclic subgroup generated by each element:

$$\langle 1 \rangle = \{1\},$$
$$\langle 5 \rangle = \{1, 5\},$$
$$\langle 7 \rangle = \{1, 7\},$$
$$\langle 11 \rangle = \{1, 11\},$$
$$\langle 13 \rangle = \{1, 13\},$$
$$\langle 17 \rangle = \{1, 17\},$$
$$\langle 19 \rangle = \{1, 19\},$$
$$\langle 23 \rangle = \{1, 23\}.$$

Remark This group has a large number of non-cyclic subgroups of order 4: for example,

$$\{1, 5, 7, 11\} \quad \text{and} \quad \{1, 5, 13, 17\}.$$

4.12 In $(\mathbb{Z}_{13}^*, \times_{13})$,

$$\langle 1 \rangle = \{1\},$$
$$\langle 2 \rangle = \{1, 2, 4, 8, 3, 6, 12, 11, 9, 5, 10, 7\}$$
$$= \mathbb{Z}_{13}^* = \langle 2^{-1} \rangle = \langle 7 \rangle,$$
$$\langle 3 \rangle = \{1, 3, 9\} = \langle 3^{-1} \rangle = \langle 9 \rangle,$$
$$\langle 4 \rangle = \{1, 4, 3, 12, 9, 10\} = \langle 4^{-1} \rangle = \langle 10 \rangle,$$
$$\langle 5 \rangle = \{1, 5, 12, 8\} = \langle 5^{-1} \rangle = \langle 8 \rangle,$$
$$\langle 6 \rangle = \{1, 6, 10, 8, 9, 2, 12, 7, 3, 5, 4, 11\}$$
$$= \mathbb{Z}_{13}^* = \langle 6^{-1} \rangle = \langle 11 \rangle,$$
$$\langle 12 \rangle = \{1, 12\}.$$

Hence $(\mathbb{Z}_{13}^*, \times_{13})$ is cyclic, with generators 2, 6, 7 and 11. By Theorem 4.4, all the subgroups of $(\mathbb{Z}_{13}^*, \times_{13})$ are cyclic, so the above list gives all the subgroups of this group.

Index